Introduction

This revision and classroom companion is matched to the new single award AQA GCSE Physics specification. It provides full coverage of the three units of substantive content (scientific explanations and evidence) and the procedural content, How Science Works, which is explained thoroughly on pages 4–11.

For each sub-section of substantive content, the AQA GCSE Physics specification identifies activities you should be able to complete using your skills, knowledge and understanding of how science works. Many of these activities focus on issues that highlight the role of science in society and the impact it has on our lives, and require you to evaluate information, develop arguments and draw conclusions.

The activities are dealt with in this guide on the How Science Works pages (picked out with shaded backgrounds), which are integrated into the three main units. The points identified on these pages are designed to provide a starting point, from which you can begin to develop your own conclusions. They are not meant to be definitive or prescriptive.

At the end of each unit you will find a page of exam-style questions complete with model answers, to help you understand what is expected of you in the exam. Unit 1 features a combination of multiple choice questions and longer structured questions to reflect the different methods of assessment; Unit 2 and Unit 3 just have structured questions.

At the end of each unit there is also a page of key words and their meanings. These pages can be used as checklists to help you with your revision. Make sure you are familiar with all the words listed and understand their meanings and relevance – they are central to your understanding of the material in that unit!

HT Throughout this volume, material which is higher tier only appears in a coloured box, and can be easily identified by the symbol HT.

This guide is intended as a source of first-rate study material for GCSE students but it is our hope that it will also ease the burden of over-worked science departments.

How to Use this Revision and Classroom Companion

This guide contains everything you need to know in a user friendly format. In certain places we have included slightly more than the specification suggests you need to know to aid understanding.

Don't just read the guide: learn actively! Constantly test yourself without looking at the text.

Jot down anything you think will help you to remember – no matter how trivial it may seem.

Contents

Contents

The numbers in brackets correspond to the reference numbers on the AQA GCSE Physics specification.

How Science Works

The new AQA GCSE Physics specification incorporates two types of content:

- **Science Content** (example shown opposite)
  This is all the scientific explanations and evidence that you will need to be able to recall in your exams (objective tests or written exams). It is covered on pages 12–100 of the revision guide.
- **How Science Works** (example shown opposite)
  This is a set of key concepts, relevant to all areas of science. It is concerned with how scientific evidence is obtained and the effect it has on society. More specifically, it covers…
  - the relationship between scientific evidence and scientific explanations and theories
  - the practices and procedures used to collect scientific evidence
  - the reliability and validity of scientific evidence
  - the role of science in society and the impact it has on our lives
  - how decisions are made about the use of science and technology in different situations, and the factors affecting these decisions.

Because they are interlinked, your teacher will have taught the two types of content together in your science lessons. Likewise, the questions on your exam papers are likely to combine elements from both types of content, i.e. to answer them, you will need to recall the relevant scientific facts *and* draw upon your knowledge of how science works.

The key concepts from How Science Works are summarised in this section of the revision guide. You should be familiar with all of them, especially the practices and procedures used to collect scientific data (from all your practical investigations). But make sure you work through them all. Make a note if there is anything you are unsure about and then ask your teacher for clarification.

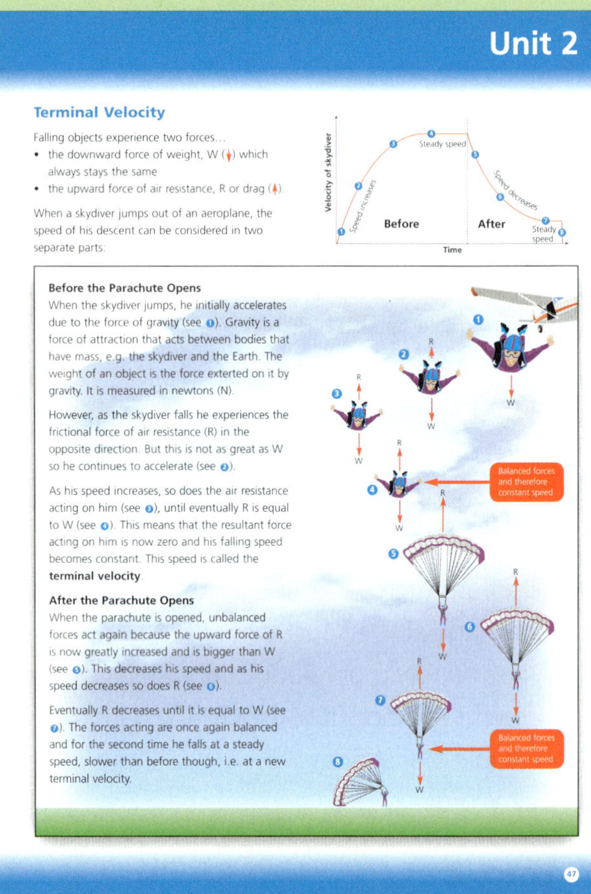

Unit 2

Terminal Velocity

Falling objects experience two forces…
- the downward force of weight, W (↓) which always stays the same
- the upward force of air resistance, R or drag (↑).

When a skydiver jumps out of an aeroplane, the speed of his descent can be considered in two separate parts:

Before the Parachute Opens
When the skydiver jumps, he initially accelerates due to the force of gravity (see ❶). Gravity is a force of attraction that acts between bodies that have mass, e.g. the skydiver and the Earth. The weight of an object is the force exterted on it by gravity. It is measured in newtons (N).

However, as the skydiver falls he experiences the frictional force of air resistance (R) in the opposite direction. But this is not as great as W so he continues to accelerate (see ❷).

As his speed increases, so does the air resistance acting on him (see ❸), until eventually R is equal to W (see ❹). This means that the resultant force acting on him is now zero and his falling speed becomes constant. This speed is called the **terminal velocity**.

After the Parachute Opens
When the parachute is opened, unbalanced forces act again because the upward force of R is now greatly increased and is bigger than W (see ❺). This decreases his speed and as his speed decreases so does R (see ❻).

Eventually R decreases until it is equal to W (see ❼). The forces acting are once again balanced and for the second time he falls at a steady speed, slower than before though, i.e. at a new terminal velocity.

47

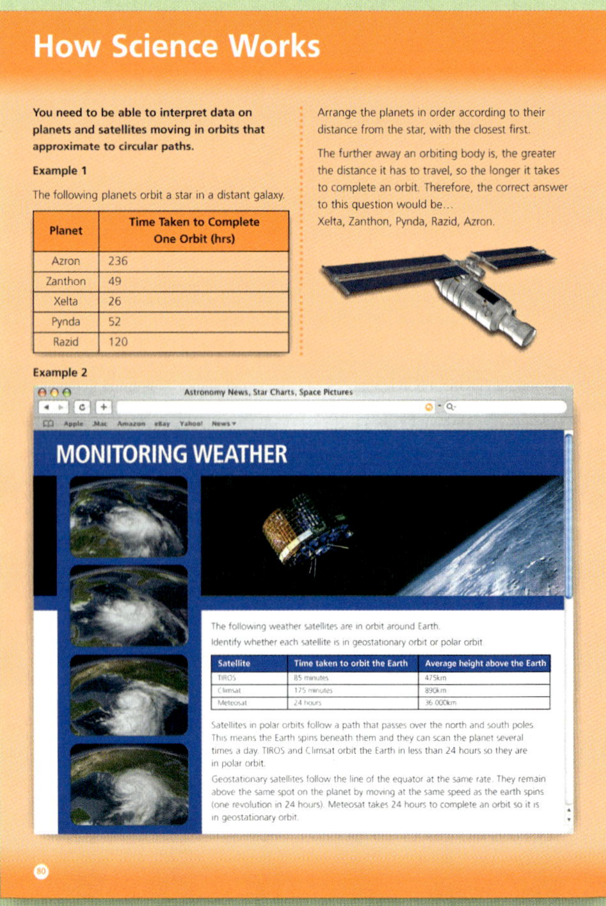

How Science Works

You need to be able to interpret data on planets and satellites moving in orbits that approximate to circular paths.

Example 1

The following planets orbit a star in a distant galaxy.

Planet	Time Taken to Complete One Orbit (hrs)
Azron	236
Zanthon	49
Xelta	26
Pynda	52
Razid	120

Arrange the planets in order according to their distance from the star, with the closest first.

The further away an orbiting body is, the greater the distance it has to travel, so the longer it takes to complete an orbit. Therefore, the correct answer to this question would be…
Xelta, Zanthon, Pynda, Razid, Azron.

Example 2

MONITORING WEATHER

The following weather satellites are in orbit around Earth.
Identify whether each satellite is in geostationary orbit or polar orbit.

Satellite	Time taken to orbit the Earth	Average height above the Earth
TIROS	85 minutes	475km
Climsat	175 minutes	890km
Meteosat	24 hours	36 000km

Satellites in polar orbits follow a path that passes over the north and south poles. This means the Earth spins beneath them and they can scan the planet several times a day. TIROS and Climsat orbit the Earth in less than 24 hours so they are in polar orbit.

Geostationary satellites follow the line of the equator at the same rate. They remain above the same spot on the planet by moving at the same speed as the earth spins (one revolution in 24 hours). Meteosat takes 24 hours to complete an orbit so it is in geostationary orbit.

80

How to Use this Revision Guide

The AQA GCSE Physics specification includes activities for each sub-section of science content. The activities require you to apply your knowledge of how science works to help develop your skills when it comes to evaluating information, developing arguments and drawing conclusions.

These activities are dealt with on the How Science Works pages (on a tinted background) throughout the revision guide. Make sure you work through them all, as questions relating to the skills, ideas and issues covered on these pages could easily come up in the exam. Bear in mind that these pages are designed to provide a starting point from which you can begin to develop your own ideas and conclusions. They are not meant to be definitive or prescriptive.

Practical tips on how to evaluate information are included in this section, on page 11.

What is the Purpose of Science?

Science attempts to explain the world we live in. The role of a scientist is to collect evidence through investigations to...

- explain phenomena (i.e. explain how and why something happens)
- solve problems.

Scientific knowledge and understanding can lead to the development of new technologies (e.g. in medicine and industry) which have a huge impact on society and the environment.

Scientific Evidence

The purpose of evidence is to provide facts which answer a specific question, and therefore support or disprove an idea or theory. In science, evidence is often based on data that has been collected by making observations and measurements.

To allow scientists to reach appropriate conclusions, evidence must be...

- **reliable**, i.e. it must be reproducible by others and therefore be trustworthy
- **valid**, i.e. it must be reliable *and* it must answer the question.

N.B. If data is not reliable, it cannot be valid.

To ensure scientific evidence is reliable and valid, scientists employ a range of ideas and practices which relate to...

1. **observations** – how we observe the world
2. **investigations** – designing investigations so that patterns and relationships can be identified
3. **measurements** – making measurements by selecting and using instruments effectively
4. **presenting data** – presenting and representing data
5. **conclusions** – identifying patterns and relationships and drawing suitable conclusions.

These five key ideas are covered in more detail on the following pages.

How Science Works

Observations

Most scientific investigations begin with an observation, i.e. a scientist observes an event or phenomenon and decides to find out more about how and why it happens.

The first step is to develop a **hypothesis**, i.e. to *suggest* an explanation for the phenomenon. Hypotheses normally propose a relationship between two or more variables (factors that change). They are based on careful observations and existing scientific knowledge, and often include a bit of creative thinking.

The hypothesis is used to make a prediction, which can be tested through scientific investigation. The data collected during the investigation might support the hypothesis, show it to be untrue, or lead to the development of a new hypothesis.

Example

A biologist **observes** that freshwater shrimp are only found in certain parts of a stream.

He uses current scientific knowledge of shrimp behaviour and water flow to develop a **hypothesis**, which relates the distribution of shrimp (first variable) to the rate of water flow (second variable).

Based on this hypothesis, the biologist **predicts** that shrimp can only be found in areas of the stream where the flow rate is beneath a certain value.

The prediction is **investigated** through a survey, which looks for the presence of shrimp in different parts of the stream, representing a range of different flow rates.

The **data** shows that shrimp are only present in parts of the stream where the flow rate is below a certain value (i.e. it supports the hypothesis). However, it also shows that shrimp are not *always* present in parts of the stream where the flow rate is below this value.

As a result, the biologist realises there must be another factor affecting the distribution of shrimp. So, he **refines his hypothesis**, to relate the distribution of shrimp (first variable) to the concentration of oxygen in the water (second variable) in parts of the stream where there is a slow flow rate.

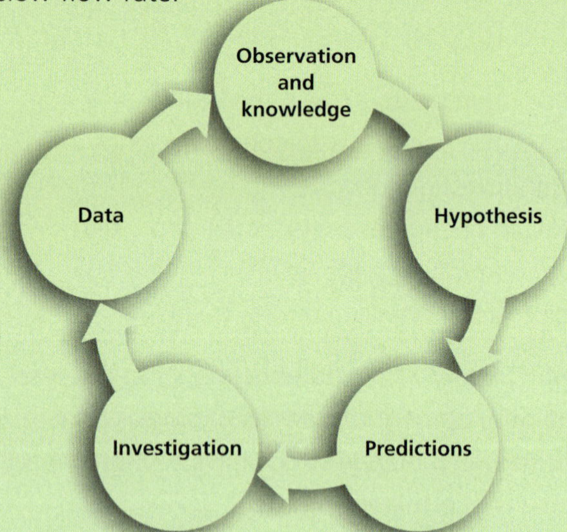

If new observations or data do not match existing explanations or theories, e.g. if unexpected behaviour is displayed, they need to be checked for reliability and validity.

In some cases, it turns out that the new observations and data are valid, so existing theories and explanations have to be revised or amended. This is how scientific knowledge gradually grows and develops.

Investigations

An investigation involves collecting data to try to determine whether there is a relationship between two variables. A variable is any factor that can take different values (i.e. change). In an investigation you have two variables:

- **independent variable**, which is controlled or known by the person carrying out the investigation. In the shrimp example on page 6, the independent variable is the flow rate of the water.
- **dependent variable**, which is measured each time a change is made to the independent variable, to see if it also changes. In the shrimp example on page 6, the dependent variable is the distribution of shrimp (i.e. whether shrimp are present or not).

Variables can have different types of values:

- **continuous variables** – can take any numerical values. These are usually measurements, e.g. temperature or height.
- **discrete variables** – can only take whole-number values. These are usually quantities, e.g. the number of shrimp in a population.
- **ordered variables** – have relative values, e.g. small, medium or large.
- **categoric variables** – have a limited number of specific values, e.g. the different breeds of dog: dalmatian, cocker spaniel, labrador, etc.

Numerical values tend to be more powerful and informative than ordered variables and categoric variables.

An investigation tries to establish whether an observed link between two variables is...

- **causal** – a change in one variable causes a change in the other, e.g. in a chemical reaction the rate of reaction (dependent variable) increases when the temperature of the reactants (independent variable) is increased
- **due to association** – the changes in the two variables are linked by a third variable, e.g. a link between the change in pH of a stream (first variable) and a change in the number of different species found in the stream (second variable), may be the effect of a change in the concentration of atmospheric pollutants (third variable)
- **due to chance** – the change in the two variables is unrelated; it is coincidental, e.g. in the 1940s the number of deaths due to lung cancer increased as did the amount of tar being used in road construction, however, one *did not* cause the other.

Investigations (continued)

Fair Test

A fair test is one in which the only factor that can affect the dependent variable is the independent variable. Any other variables (outside variables) that could influence the results are kept the same.

This is a lot easier in the laboratory than in the field, where conditions (e.g. weather) cannot always be physically controlled. The impact of outside variables, like the weather, has to be reduced by ensuring all measurements are affected by the variable in the same way. For example, if you were investigating the effect of different fertilisers on the growth of tomato plants, all the plants would need to be grown in a place where they were subject to the same weather conditions.

If a survey is used to collect data, the impact of outside variables can be reduced by ensuring that the individuals in the sample are closely matched. For example, if you were investigating the effect of smoking on life expectancy, the individuals in the sample would all need to have a similar diet and lifestyle to ensure that those variables do not affect the results.

Control groups are often used in biological research. For example, in some drugs trials, a placebo (a dummy pill containing no medicine) is given to one group of volunteers – the control group – and the drug is given to another. By comparing the two groups, scientists can establish whether the drug (the independent variable) is the only variable affecting the volunteers and, therefore, whether it is a fair test.

Accuracy and Precision

In an investigation, the mean (average) of a set of repeated measurements is often calculated to overcome small variations and get a best estimate of the true value. Increasing the number of measurements taken will improve the accuracy and the reliability of their mean.

$$\text{Mean} = \frac{\text{Sum of all measurements}}{\text{Number of measurements}}$$

The purpose of an investigation will determine how accurate the data collected needs to be. For example, measures of blood alcohol levels must be accurate enough to determine whether a person is legally fit to drive.

The data collected must also be precise enough to form a valid conclusion, i.e. it should provide clear evidence for or against the hypothesis.

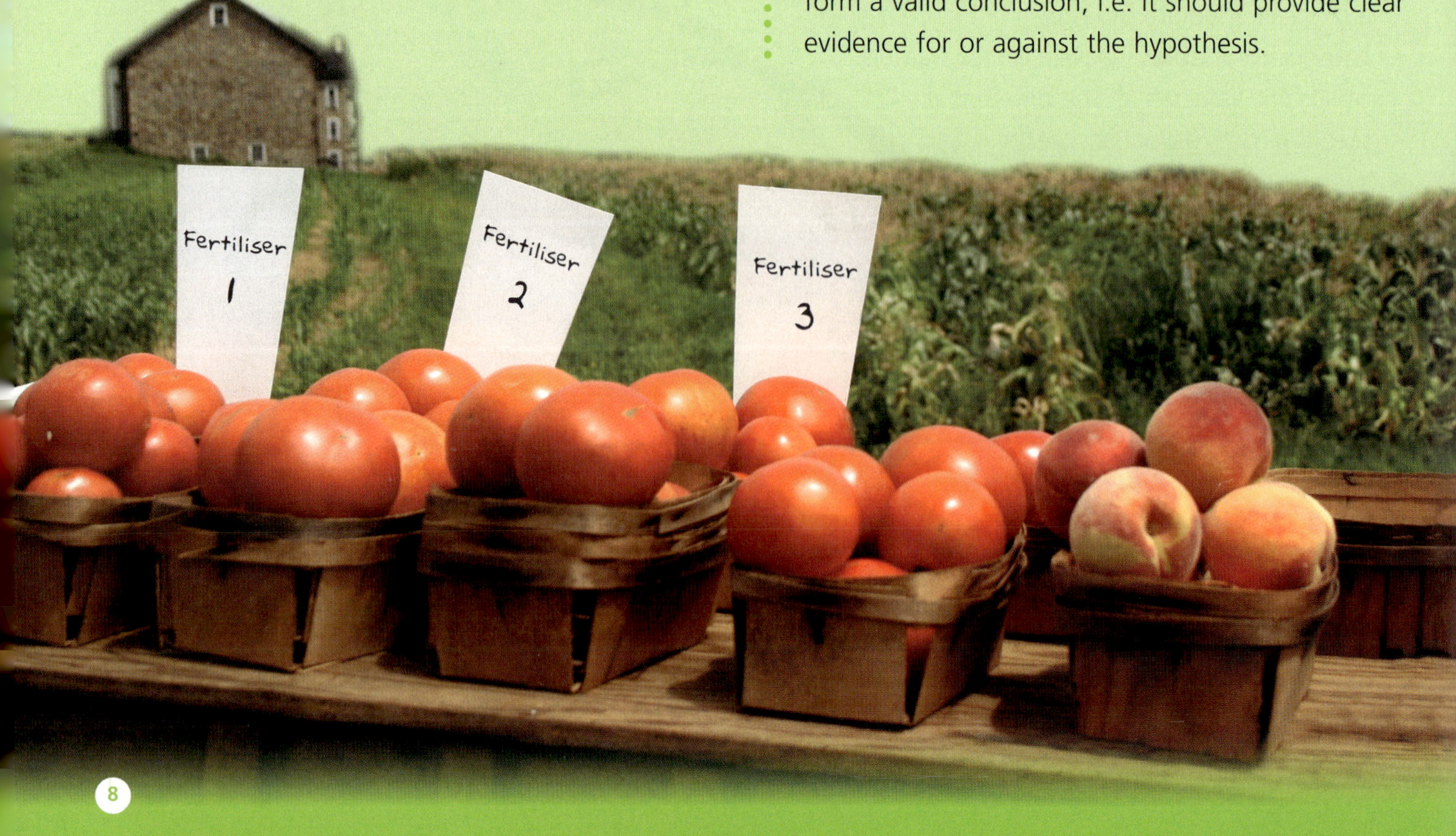

Measurements

Even if all outside variables have been controlled, there are certain factors that could still affect the reliability and validity of any measurements made:

- **The accuracy of the instruments used** – The accuracy of a measuring instrument will depend on how accurately it has been calibrated. Expensive equipment is likely to be more accurately calibrated.
- **The sensitivity of the instruments used** – The sensitivity of an instrument is determined by the smallest change in value it can detect. For example, bathroom scales are not sensitive enough to detect the changes in weight of a small baby, whereas the scales used by a midwife to monitor growth are.
- **Human error** – When making measurements, random errors can occur due to a lapse in concentration, and systematic (repeated) errors can occur if the instrument has not been calibrated properly or is repeatedly misused.

Any anomalous (irregular) values, i.e. values that fall well outside the range (the spread) of the other measurements, need to be examined to try to determine the cause. If they have been caused by an equipment failure or human error, it is common practice to ignore such values and discount them from any subsequent calculations.

Range = Maximum value − Minimum value

Presenting Data

Data is often presented in a format that makes the patterns more evident. This makes it easier to see the relationship between two variables. The relationship between variables can be linear (positive or negative) or directly proportional.

Clear presentation of data also makes it easier to identify any anomalous values.

The type of chart or graph used to present data will depend on the type of variable involved.

Tables can be used to organise data (patterns and anomalies in the data are not always obvious).

Height (cm)	127	165	149	147	155	161	154	138	145
Shoe size	5	8	5	6	5	5	6	4	5

Bar charts are used to display data when the independent variable is categoric or discrete and the dependent variable is continuous.

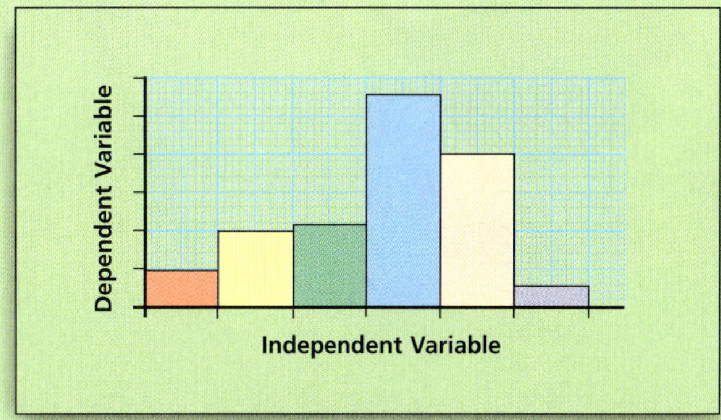

Line graphs are used to display data when both variables are continuous.

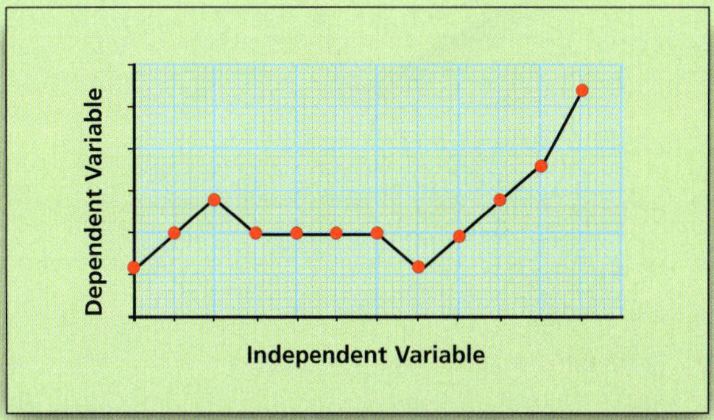

Scattergrams (or scatter diagrams) are used to show the underlying relationship between two variables. This can be made clearer by including a line of best fit.

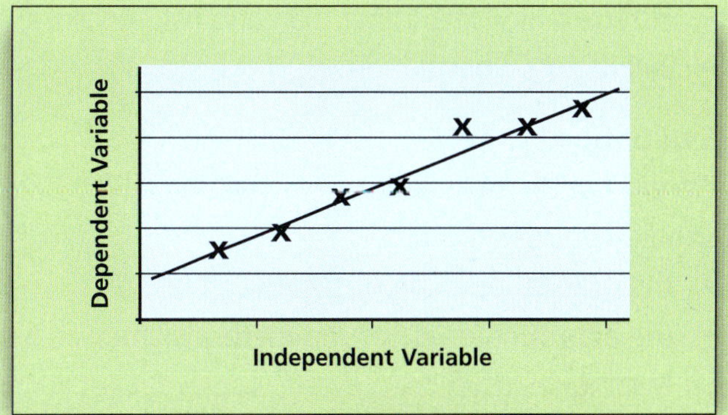

How Science Works

Conclusions

Conclusions should...

- describe the patterns and relationships between variables
- take all the data into account
- make direct reference to the original hypothesis / prediction.

Conclusions should not...

- be influenced by anything other than the data collected
- disregard any data (other than anomalous values)
- include any speculation.

Evaluation

An evaluation looks at the investigation as a whole. It should consider...

- the original purpose of the investigation
- the appropriateness of the methods and techniques used
- the reliability and validity of the data
- the validity of the conclusions (e.g. whether the original purpose was achieved).

The reliability of an investigation can be increased by...

- looking at relevant data from secondary sources
- using an alternative method to check results
- ensuring that the results can be reproduced by others.

Science and Society

Scientific understanding can lead to technological developments, which can be exploited by different groups of people for different reasons. For example, the successful development of a new drug benefits the drugs company financially and improves the quality of life for patients.

The applications of scientific and technological developments can raise certain issues. An issue is an important question that is in dispute and needs to be settled. Decisions made by individuals and society about these issues may not be based on scientific evidence alone.

Social issues are concerned with the impact on the human population of a community, city, country, or even the world.

Economic issues are concerned with money and related factors like employment and the distribution of resources. There is often an overlap between social and economic issues.

Environmental issues are concerned with the impact on the planet; its natural ecosystems and resources.

Ethical issues are concerned with what is morally right and wrong, i.e. they require a value judgement to be made about what is acceptable. As society is underpinned by a common belief system, there are certain actions that can never be justified. However, because the views of individuals are influenced by lots of different factors (e.g. faith and personal experience) there are also lots of 'grey areas'.

Evaluating Information

It is important that you can evaluate information relating to social-scientific issues. You could be asked to do this in the exam, but it will also help you to make informed decisions in life (e.g. decide whether or not to have a particular vaccination or become involved in a local recycling campaign).

When you are asked to **evaluate** information, start by making a list of the pluses and the minuses. Then work through the two lists, and for each point consider how this might impact on society. Remember, **PMI** – pluses, minuses, impact on society.

You also need to be sure that the source of information is reliable and credible. Here are some important factors to consider:

- **Opinion (personal viewpoints) –** opinions which are backed up by valid and reliable evidence carry far more weight than those based on non-scientific ideas (e.g. hearsay or urban myths).
- **Bias –** information is biased if it does not provide a balanced account; it favours a particular viewpoint. Biased information might include incomplete evidence or try to influence how you interpret the evidence. For example, a drugs company might highlight the benefits of their drugs but downplay the side effects in order to increase sales.
- **Weight of evidence –** scientific evidence can be given undue weight or dismissed too lightly due to...
  - political significance, e.g. evidence that is likely to provoke an extreme and negative reaction from the public might be downplayed
  - status (academic or professional status, experience, authority and reputation of the scientist who undertook the study), e.g. evidence is likely to be given more weight if it comes from someone who is a recognised expert in that particular field.

Limitations of Science

Science can help us in lots of ways but it cannot supply all the answers. We are still finding out about things and developing our scientific knowledge. There are some questions that we cannot answer, maybe because we do not have enough reliable and valid evidence.

There are some questions that science cannot answer at all. These tend to be questions relating to ethical issues, where beliefs and opinions are important, or to situations where we cannot collect reliable and valid scientific evidence. In other words, science can often tell us whether something *can* be done and *how* it can be done, but it cannot tell us whether it *should* be done.

Unit 1

11.1

How is heat (thermal energy) transferred and what factors affect the rate at which it is transferred?

Heat can be transferred from hotter places to colder places by three different methods: conduction, convection and radiation. To transfer heat effectively, with minimal heat loss, we need to know which method is the most efficient in a particular case. To understand this, you need to know...

- the differences between the transfer methods
- which surfaces are good absorbers or emitters of radiation
- the factors that affect the rate of heat transfer.

Conduction

Conduction is the transfer of heat energy without the substance itself moving. The structure of metals makes them good conductors of heat. As a metal becomes hotter, its tightly packed particles gain more kinetic energy and vibrate. This energy is transferred to cooler parts of the metal by delocalised electrons, which move freely through the metal, colliding with particles and other electrons.

N.B. an electron is a subatomic particle (see p.29).

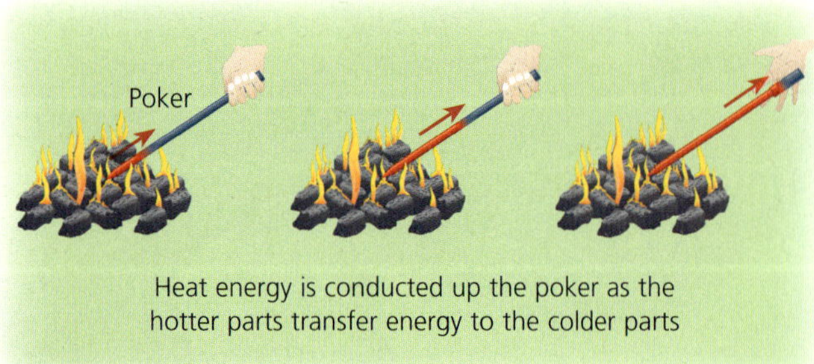

Heat energy is conducted up the poker as the hotter parts transfer energy to the colder parts

Convection

Convection is the transfer of heat energy through movement. This occurs in liquids and gases and creates convection currents.

In a liquid or gas, the particles nearest the heat source move faster causing the substance to expand and become less dense than in the colder parts. The warm liquid or gas will rise up and colder, denser liquid or gas moves into the space created (close to the heat source).

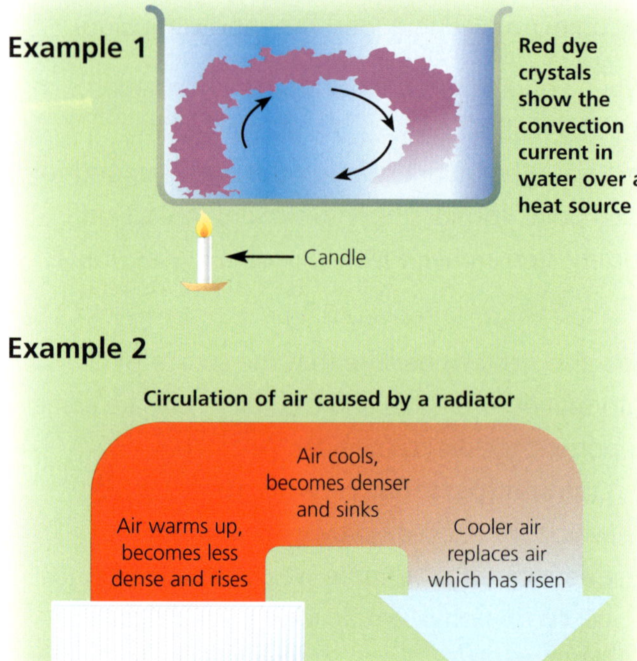

Radiation

Thermal, or infra red, **radiation** is the transfer of heat energy by electromagnetic waves; no particles of matter are involved.

All objects emit and absorb radiation. The hotter the object, the more energy it radiates. How much radiation is given out or taken in by an object depends on its surface, shape and dimensions.

An object will emit or absorb energy faster if there is a big difference in temperature between it and its surroundings. The rate of heat transfer can be slowed down by the use of insulation, which provides a barrier.

Under similar conditions, different materials transfer heat at different rates. At the same temperature...

- dark matt surfaces emit more radiation than light shiny surfaces
- dark matt surfaces absorb more radiation than light shiny surfaces.

You need to be able to evaluate ways in which heat is transferred in and out of bodies and ways in which the rates of transfer can be reduced.

Every year, people in the UK spend a lot more money on energy (gas and electricity) to heat their homes than they need to. This is because energy gets lost and wasted.

An uninsulated house can lose up to 75% of heat through the roof, walls, windows, floors and doors. Heat energy is transferred from homes into the environment by...

- **conduction** – through the walls, floor, roof and windows
- **convection** – convection currents coupled with cold draughts from gaps in doors and windows cause heat energy to rise up to the roof space where it is easily lost
- **radiation** – through the walls, roof and windows.

Arrows show percentage of total heat loss

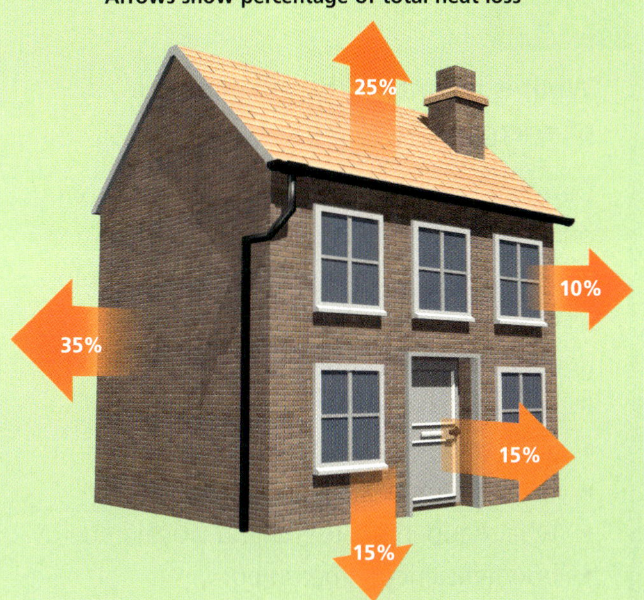

Insulating a house will help it to retain the heat in winter, and also help to keep it cool in the summer. The table below outlines how heat can be lost and how this loss can be reduced.

Where Heat is Lost	Preventative Measure	Benefits	Problems
Roof	Roof insulation – traps a layer of air between fibres or insulating material.	• Can reduce heat loss by 20–25%. • Many different methods to suit all homes.	• Needs to be laid by an expert. • Expensive.
Under doors and windows	Draught excluders – keeps as much warm air inside as possible.	• Can reduce heat loss by up to 15%. • Cheap and easy to install.	• Must make sure that air vents are not blocked – need fresh air to circulate to prevent dry rot.
Walls	Cavity wall insulation and internal thermal boards.	• Can reduce heat loss by 35%.	• Expensive.
Windows	Double glazing – traps air between two sheets of glass. Curtains – stop heat loss through convection.	• Double glazing can reduce heat loss by up to 10%. • Curtains are cheap and easy to install.	• Double glazing is expensive.
Floor	Carpets, rugs and underfloor insulation can help to stop heat loss through the floor.	• Carpets and rugs are easy to install.	• Underfloor insulation is expensive.

11.2

What is meant by the efficient use of energy?

Devices can transform (change) energy from one form into another. When energy is transformed, not all of it is transformed into useful energy; some of it is wasted. To understand energy efficiency, you need to know…

- how energy is transferred
- how energy is wasted
- what happens to useful and wasted energy
- how efficiency is measured.

Transferring and Transforming Energy

When devices transfer energy, only part of the energy is usefully **transferred** to where it is wanted and in the form that it is wanted. The remaining energy is **transformed** in a non-useful way, mainly as heat energy that goes into the surroundings, and is known as **wasted** energy.

The **wasted** and the **useful** energy are eventually transferred to their surroundings which become warmer.

No energy is destroyed or created, it is just changed into a different form (transformed). However, the energy becomes increasingly spread out, so it is difficult for any further useful energy transfers to occur.

Efficiency of Devices

The greater the proportion of energy that is usefully transformed, the more **efficient** we say the device is. Opposite are examples of the efficiency of two devices.

- A car engine is 20% efficient. Much more energy is wasted (in heat and sound), than is transformed into useful kinetic energy.
- A microwave is 60% efficient. More energy is transformed into useful heat, light and kinetic energy than is wasted (as heat and sound).

Only a quarter of the energy supplied to a television is usefully transformed into light and sound. Therefore it is only 25% efficient.

You need to be able to describe the intended energy transfers / transformations and the main energy wastage that occur with a range of devices, and calculate the efficiency of a device using:

$$\text{Efficiency} = \frac{\text{Useful energy transferred by device}}{\text{Total energy supplied to device}}$$

Some examples are shown in the table below.

Name of Device and Intended Energy Transfer	Energy In	Energy Out		Efficiency
		Useful	Wasted	
Standard light bulb – electrical energy to light energy	100 joules/sec	Light: 20 joules/sec	Heat: 80 joules/sec	$\frac{20}{100}$ x 100% = **20%**
Low energy light bulb – electrical energy to light energy	25 joules/sec	Light: 20 joules/sec	Heat: 5 joules/sec	$\frac{20}{25}$ x 100% = **80%**
Kettle – electrical energy to heat energy	2000 joules/sec	Heat (in water): 1800 joules/sec	Heat: 100 joules/sec Sound: 100 joules/sec	$\frac{1800}{2000}$ x 100% = **90%**
Electric motor – electrical energy to kinetic energy	500 joules/sec	Kinetic: 300 joules/sec	Heat: 100 joules/sec Sound: 100 joules/sec	$\frac{300}{500}$ x 100% = **60%**

You need to be able to evaluate the effectiveness and cost effectiveness of methods used to reduce energy consumption.

Method	Benefits	Problems
Switching lights off when leaving a room.	• Easy and simple way to reduce energy consumption.	• People forget or do not like being in a dark house.
Energy-efficient light bulbs.	• Use less power so cost less to run. • Less wasted energy. • Last longer than standard bulbs.	• More expensive to buy.
Using electrical equipment during the night, e.g. washing machine.	• Cheaper time of day for using electricity.	• Can be noisy and keep people awake.
More efficient tumble driers, or letting clothes dry naturally.	• Driers with a sensor stop automatically when the clothes are dry, which saves energy.	• New driers are expensive to buy. • Cannot hang clothes out to dry if it is raining.
Tankless water heater.	• Water is heated when needed so less energy is used to heat unnecessary water, and keep it heated.	• Some units do not have enough power to supply to more than one tap at a time.

11.3

Why are electrical devices so useful?

Electrical energy can be transferred easily across large distances and therefore is very suitable to use in our homes and industries. To understand this, you need to know...

- what energy transformations electrical devices bring about
- how energy and power are measured
- how electricity is transferred
- how transformers are used
- what happens when voltage is increased.

Energy Transformation

Most of the energy transferred to homes and industry is electrical energy because it is easily transformed to...

- heat (thermal) energy, e.g. an electric fire
- light energy, e.g. a lamp
- sound energy, e.g. stereo speakers
- movement (kinetic) energy, e.g. an electric whisk.

The **power** of an appliance is measured in **watts (W)** or **kilowatts (kW)**, and **energy** is normally measured in **joules (J)**.

The amount of energy transformed by an electrical appliance depends on...

- how long the appliance is switched on
- how fast the appliance can transform energy.

Energy Transfer

Electricity generated at power stations is distributed to homes, schools and factories all over the country by a network of cables called the National Grid.

Transformers are used to change the voltage of the alternating current supply before and after it is transmitted through the National Grid.

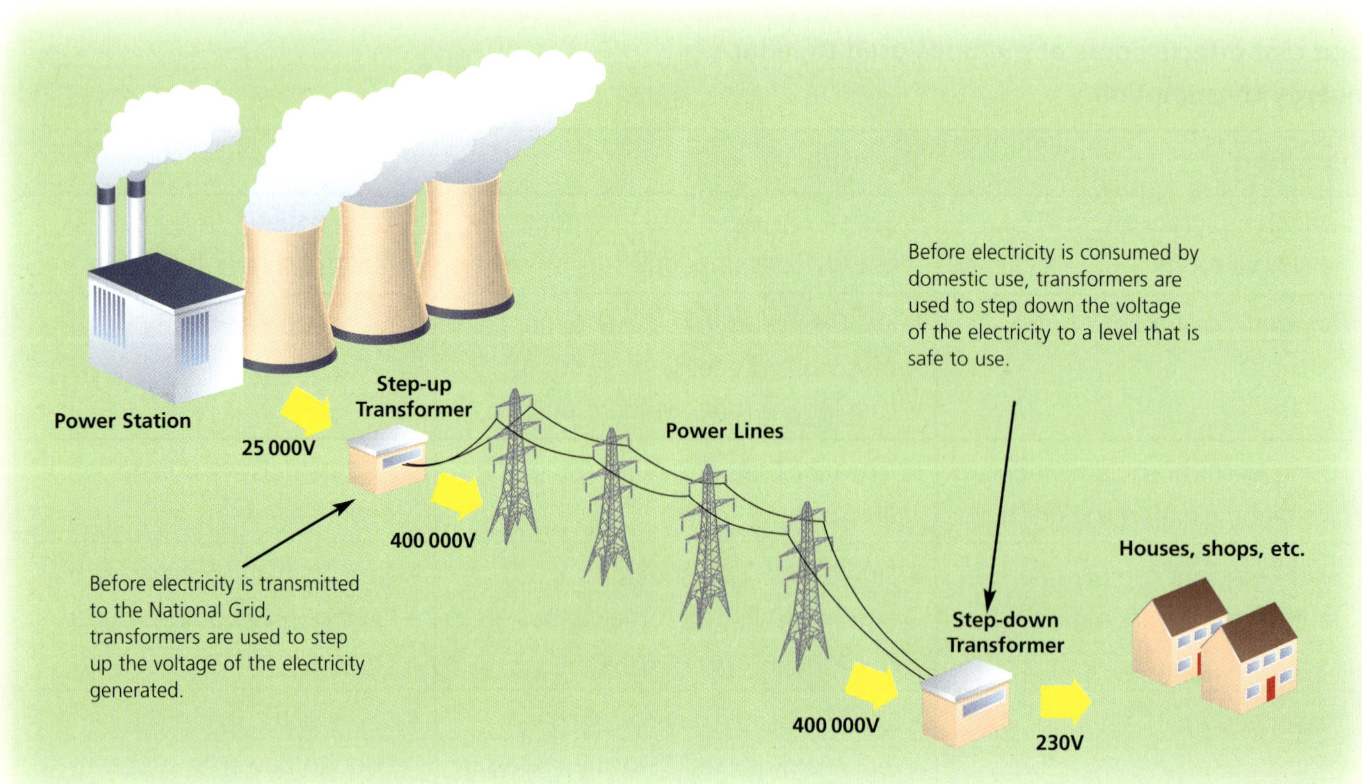

Reducing Energy Loss During Transmission

The higher the current that passes through a wire, the greater the amount of energy that is lost as heat from the wire. So we need to transmit as low a current as possible through the power lines.

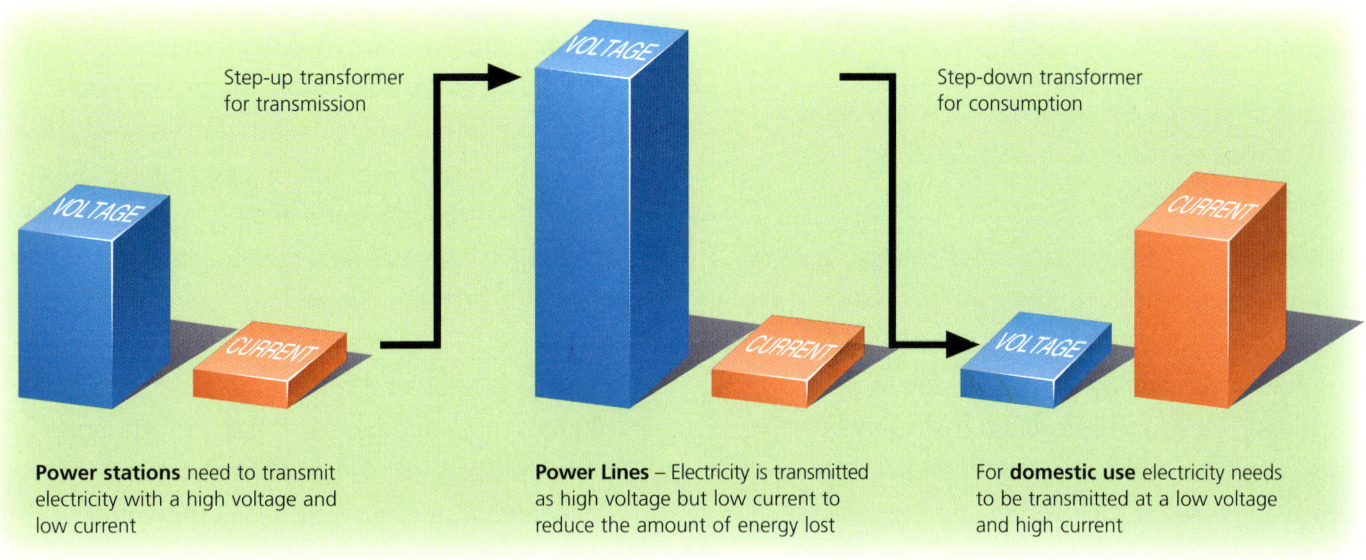

Calculating the Amount and Cost of Energy Transferred

The **amount of energy** transferred from the mains can be calculated using the following equation...

Energy transferred (kilowatt-hour, kWh)	=	Power (kilowatt, kW)	x	Time (hour, h)

The **cost of energy** transferred from the mains can then be calculated using the following equation...

Total cost	=	Number of kilowatt-hours	x	Cost per kilowatt-hour

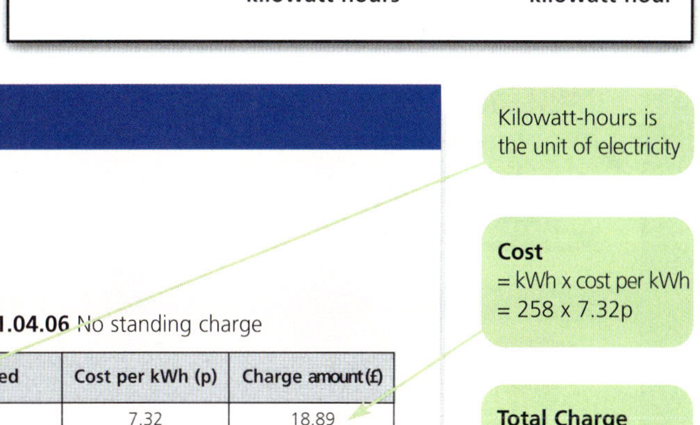

REB Regional **Electricity** Board

Mr R. Jones
273 Dove Street
Southampton
SW15 WFK

Electricity Statement. Period: 01.01.06 – 01.04.06 No standing charge

Present reading	Previous reading	kWh used	Cost per kWh (p)	Charge amount (£)
12898 (economy)	12640 (economy)	258	7.32	18.89
30803	30332	471	9.45	44.51
			Total VAT exclusive charges	63.40
			VAT at 5%	2.54
			Total charges including VAT	65.94

Regional Electricity Board, Anchor House, Ingleby Street, Southampton SW15 TNE **Telephone:** 01445 680180 **Fax:** 01445 680180 **Email:** info@reb.co.uk **Web:** www.reb.co.uk

Economy reading – electricity used during the night

Normal reading – electricity used during the daytime

kWh calculated by subtracting present reading from previous reading

Kilowatt-hours is the unit of electricity

Cost
= kWh x cost per kWh
= 258 x 7.32p

Total Charge
= economy charge + normal charge

VAT = 63.40 x $\frac{5}{100}$

How Science Works

You need to be able to compare and contrast different electrical devices that can be used for a particular application, and calculate the amount of energy transferred from the mains using...

Energy transferred (kilowatt-hour, kWh)	=	Power (kilowatt, kW)	x	Time (hour, h)

Example

Mr and Mrs Jones are shopping for a new shower for their house. They see three different showers advertised (see below).

They work out the advantages and disadvantages of each model in order to decide which one will be the most suitable for their needs. They also need to work out how much energy will be needed to power each one. They have calculated that every morning they each have a shower lasting, on average, 12 minutes, whilst their young son, Nick, has a shorter shower lasting 6 minutes (total time = 12 + 12 + 6 = 30 mins or 0.5 hours).

Using the formula they calculate how much energy will need to be transferred from the mains to power each shower:

Power x Time = Energy
7kW: 7 x 0.5 = 3.5kWh
8.5kW: 8.5 x 0.5 = 4.25kWh
9.5kW: 9.5 x 0.5 = 4.75kWh

To compare the showers they work out...

- the 7kW shower will require the least amount of energy from the mains, but it does have a less powerful spray, fewer options and needs to be connected to a hot water tank
- the 8kW shower will require more energy than the 7kW shower, but is has a more powerful spray. It also needs to be connected to the hot water tank
- the 9.5kW shower is more expensive to buy and requires more energy, but it has a much more powerful spray, more functions and can be connected to a cold water tank. This will be more useful for a family as they won't have to wait for the water to heat up.

11.4

How should we generate the electricity we need?

Various energy sources can be used to generate the electricity that we need. We need to understand the benefits and costs associated with each method to select the best one to use in a particular situation. To understand this, you need to know...

- how energy is produced
- about renewable and non-renewable energy sources
- the advantages and disadvantages of using different energy sources.

Fuels are substances which release useful amounts of energy when they are burned.

Non-renewable Energy Sources

Coal, oil and gas are energy sources which are formed over millions of years from the remains of plants and animals. They are called **fossil fuels** and are responsible for most of the energy that we use. However, because they cannot be replaced within a lifetime, they will eventually run out. They are therefore called **non-renewable** energy sources.

Non-renewable Energy Sources (Fossil Fuels)

Nuclear fuels such as uranium and plutonium are also non-renewable. Nuclear fission is the release of nuclear particles that collide with other atoms causing a chain reaction that generates huge amounts of heat energy. However, nuclear fuel is not burnt like coal, oil or gas to release energy and is not classed as a fossil fuel.

Generating Electricity from Non-renewable Energy Sources

In power stations, fossil fuels and wood are burnt to release heat energy, which boils water to produce steam. This steam drives turbines which are attached to an electrical generator.

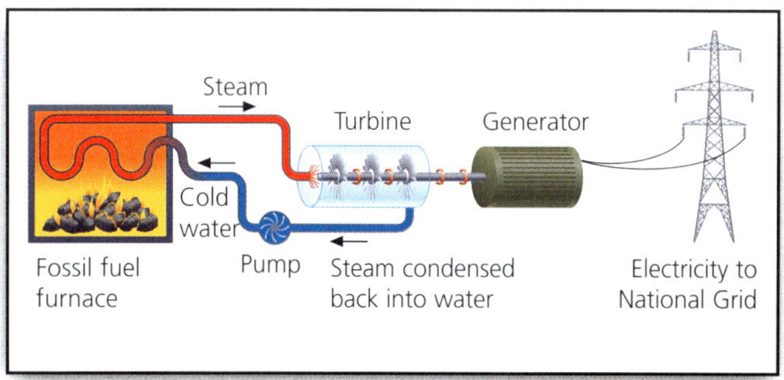

Nuclear fuel is used to generate electricity in a similar way. A reactor is used to generate heat by nuclear fission. A heat exchanger is used to transfer the heat energy from the reactor to the water which turns to steam and drives the turbines.

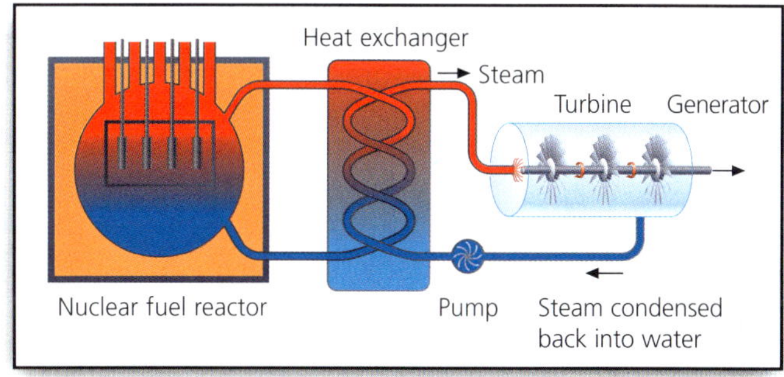

Wood

Although burnt for energy, wood is not a fossil fuel, nor is it non-renewable. It is classed as a **renewable** energy source since trees can be grown relatively quickly to replace those which are burnt to provide energy for heating.

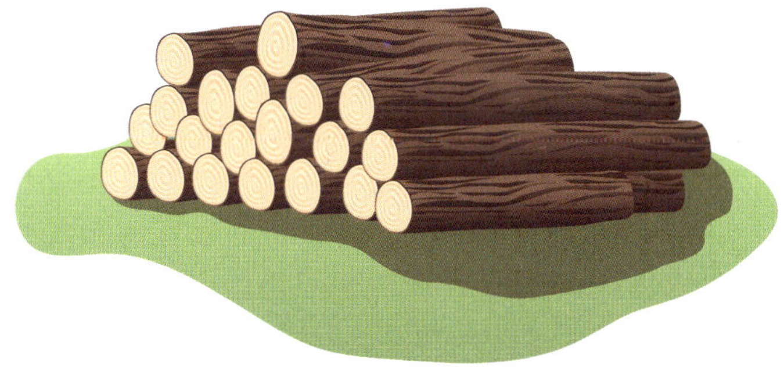

Comparing Non-renewable Sources Of Energy

The energy sources below are used to provide most of the electricity we need in this country through power stations. Some of the advantages and disadvantages of each one are listed below.

Source	Advantages	Disadvantages
Coal	• Relatively cheap and easy to obtain. • Coal-fired power stations are flexible in meeting demand and have a quicker start-up time than their nuclear equivalents. • Estimates suggest that there may be over a century's worth of coal left.	• Burning produces carbon dioxide (CO_2) and sulfur dioxide (SO_2). • Produces more CO_2, per unit of energy than oil or gas does. (CO_2 causes global warming.) • SO_2 causes acid rain unless the sulfur is removed before burning or the SO_2 is removed from the waste gases. Both of these add to the cost.
Oil	• Enough oil left for the short to medium term. • Relatively easy to find, though the price is variable. • Oil-fired power stations are flexible in meeting demand and have a quicker start-up time than both nuclear-powered and coal-fired reactors.	• Burning produces CO_2 and SO_2. (CO_2 causes global warming, SO_2 causes acid rain.) • Produces more CO_2 than gas per unit of energy. • Often carried between continents on tankers leading to the risk of spillage and pollution.
Gas	• Enough natural gas left for the short to medium term. • Can be found as easily as oil. • No SO_2 is produced. • Gas-fired power stations are flexible in meeting demand and have a quicker start-up time than nuclear, coal and oil-fired reactors.	• Burning produces CO_2, although it produces less than coal and oil, per unit of energy. (CO_2 causes global warming.) • Expensive pipelines and networks are often required to transport it to the point of use.
Nuclear	• Cost and rate of fuel is relatively low. • Can be situated in sparsely populated areas. • Nuclear power stations are flexible in meeting demand. • No CO_2 or SO_2 produced.	• Although there is very little escape of radioactive material in normal use, radioactive waste can stay dangerously radioactive for thousands of years and safe storage is expensive. • Building and de-commissioning is costly. • Longest comparative start-up time.

Summary

Advantages	Disadvantages
• Produce huge amounts of energy. • Reliable. • Flexible in meeting demand. • Do not take up much space (relatively).	• Pollute the environment. • Cause global warming and acid rain. • Will eventually run out. • Fuels often have to be transported over long distances.

Renewable Energy Sources

Renewable energy sources are those that will not run out because they are continually being replaced. Many of them are 'powered' by the Sun or Moon. The gravitational pull of the Moon creates tides and the Sun causes...

- evaporation which results in rain and flowing water
- convection currents, which result in winds, which in turn create waves.

Generating Electricity from Renewable Energy Sources

Renewable energy sources can be used to drive turbines or generators directly. In other words, nothing needs to be burnt to produce heat.

The table below shows the most common methods of generating energy from renewable energy sources.

	**Wind Turbines** Wind can be used to drive huge turbines, which, in turn, drive generators. Wind turbines are usually positioned on the top of hills so they are exposed to as much wind as possible. 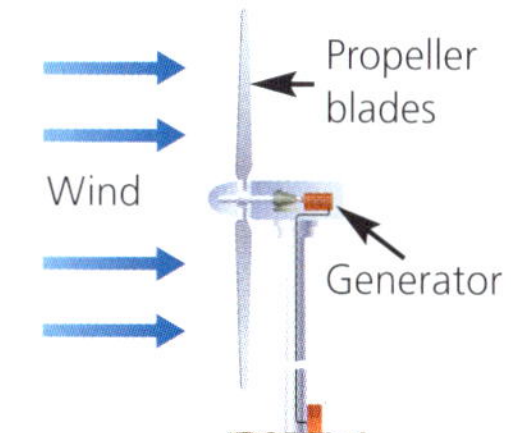
	**Solar Cells and Panels** Solar cells and panels are made of a semiconductor material (usually silicon) which captures the light energy and transforms it into electrical energy. 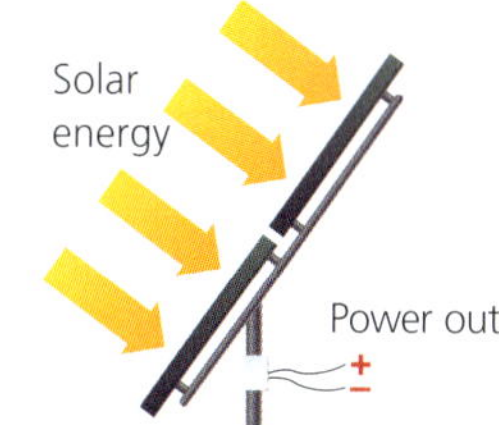
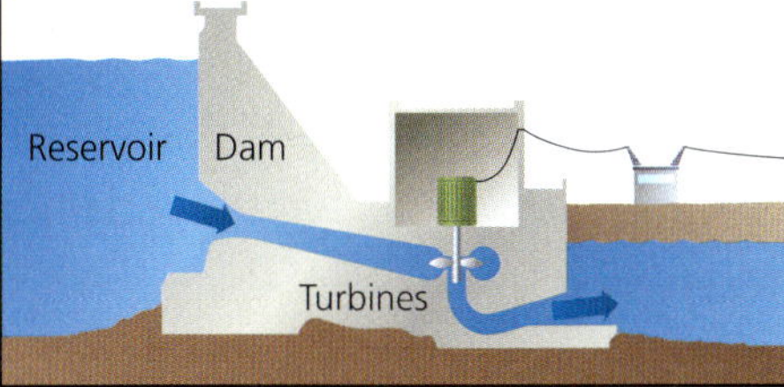	**Hydro-electric Dam** Water that is stored in a reservoir above the power station is allowed to flow down through the pipes to drive the turbines, which produces a lot of power.
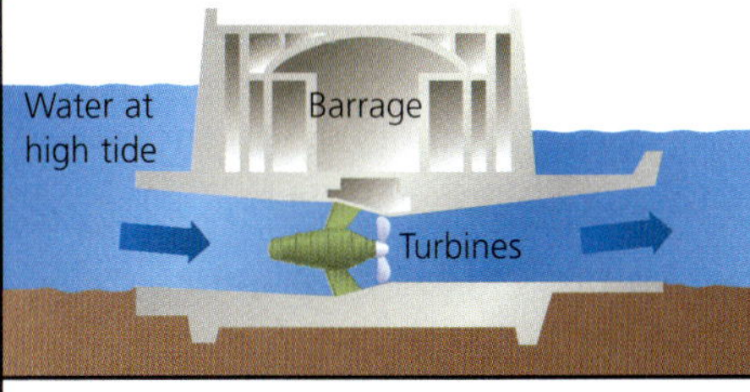	**Tidal Barrage** As the tide comes in, water flows freely through a valve in the barrage. This water then becomes trapped. At low tide, the water is released from behind the barrage through a gap which has a turbine in it. This drives a generator.
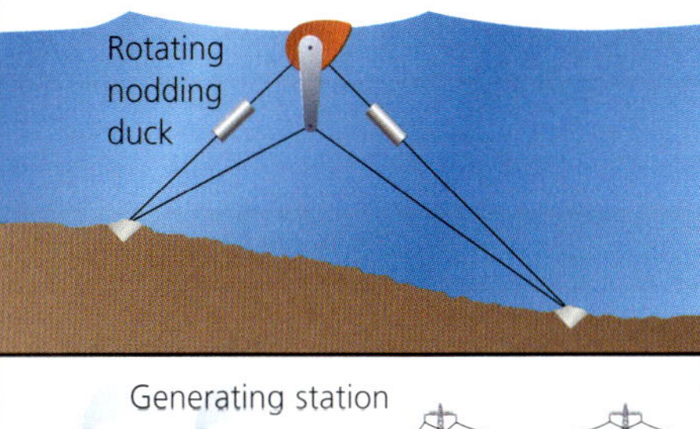	**Nodding Duck** Nodding ducks are found in the sea. The motion of the waves makes the 'ducks' rock and this movement is translated into a rotary movement which, in turn, drives a generator.
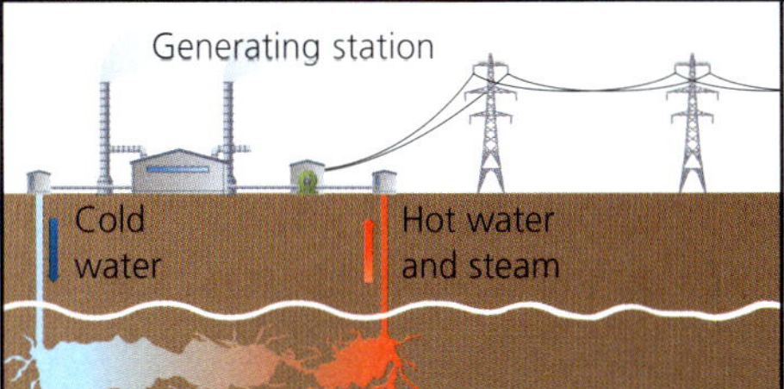	**Geothermal** In some volcanic areas, hot water and steam rise naturally to the surface, having been heated up by the decay of radioactive substances (e.g. uranium) within the Earth. This steam can be used directly to drive turbines which, in turn, drive generators.

Comparing Renewable Sources of Energy

The energy sources listed below use modern technology to provide us with a clean, safe alternative source of energy. Some of their advantages and disadvantages are given.

Source	Advantages	Disadvantages
Wind	• No fuel and little maintenance required. • No pollutant gases produced. • Once built they provide 'free' energy when the wind is blowing. • Can be built offshore.	• Need a lot to produce a sizable amount of electricity, which means noise and visual pollution. • Electricity output depends entirely on the strength of the wind. • Not very flexible in meeting demand unless the energy is stored. • Capital outlay can be high to build them.
Tidal and Waves	• No fuel required. • No pollutant gases produced. • Once built they provide 'free' energy. • Barrage water can be released when demand for electricity is high.	• Tidal barrages, across estuaries, are unsightly, a hazard to shipping, and destroy the habitats of wading birds, etc. • Daily variations of tides and waves affect output. • High initial capital outlay to build them.
Hydro-electric	• No fuel required unless storing energy to meet future demand. • Fast start-up time to meet growing demand. • Produce large amounts of clean, reliable electricity. • No pollutant gases produced. • Water can be pumped back up to the reservoir when demand for electricity is low, e.g. in the night.	• Location is critical and often involves damming upland valleys which means flooding farms, forests and natural habitats. • To achieve a net output (aside from pumping) there must be adequate rainfall in the region where the reservoir is. • Very high initial capital outlay (though worth the investment in the end).
Solar	• Ideal for producing electricity in remote locations. • Excellent energy source for small amounts. • Produces free, clean electricity. • No pollutant gases produced.	• Dependent on the intensity of light; more useful in sunny places. • High cost per unit of electricity produced compared to all other sources, except non-rechargeable batteries.

Summary

Advantages	Disadvantages
• No fuel costs during operation. • No chemical pollution. • Often low maintenance. • Do not contribute to global warming or produce acid rain.	• With the exception of hydroelectric, they produce small amounts of electricity. • Take up lots of space and are unsightly. • Unreliable (apart from hydroelectric), depend on the weather and cannot guarantee supply on demand. • High initial capital outlay.

You need to be able to compare and contrast the particular advantages and disadvantages of using different energy sources to generate electricity.

Example

The Lonsdale News, Tuesday June 13 2006 12

Power to the People

The little village of Bukestead was in uproar yesterday after proposed plans were unveiled for a nuclear power station to be built just 10 miles away from the village on the top of the moor. The local area is well known for its beautiful countryside and is a carefully controlled breeding area for many birds. However, increasing demand for electricity means that new electricity generating plants need to be built and this site is considered to be most suitable. Its isolated location and the large number of jobs it will create, have made nuclear power a popular choice with some people.

However, opinion is strongly divided as to the best use of the land, with renewable energy options being cited by many as a better alternative.

Join the debate. Write to your local MP or councillor and have your say.

6 Moorland Road
Bukestead
Derbyshire

Mrs J Smith (MP)
Derby Council Offices
134 High Street
Derby

19th June

Dear Mrs Smith

I was appalled to read about the proposed nuclear power station. The blight on the countryside, the traffic and the horror of a nuclear leak would be unthinkable! What would happen to the local countryside, farms, and the breeding grounds for migrating birds that has been so carefully built up? The local roads would be constantly clogged up with tankers travelling to and from the plant, some carrying highly dangerous waste for disposal. Not to mention the expense of building the site!

This land would be ideal for a farm of wind turbines. The constant strong winds on the hill, coupled with adequate space for 10 to 15 turbines, would be sufficient to produce a sizable amount of electricity. There would be no pollution, no fuel and little maintenance required, just clean, free energy. I hope you will consider the alternatives.

Yours sincerely,

Anne Pentwhistle

(A very concerned local resident)

26 Hill View
Bukestead
Derbyshire

Mrs J Smith (MP)
Derby Council Offices
134 High Street
Derby

17th June

Dear Mrs Smith

I am writing in support of the proposed nuclear power station.

The isolated position on the hill would be ideal. Nuclear leaks are so uncommon that no real threat would be posed and the local economy would greatly benefit from the number of jobs created.

I know that reasons are being cited for the use of 'green' electricity – big solar panels, huge wind turbine farms and the like. These are unfeasible solutions; they could never fulfil the demand, and they themselves would visually pollute the countryside. Everyone says, 'not on my door step', but nuclear is flexible, non-pollutant and will provide the electricity needed for the area.

I whole-heartedly support your plans.

Yours sincerely,

A. Pollard

Local resident

11.5

What are the uses and hazards of the waves that form the electromagnetic spectrum?

The uses and hazards of electromagnetic radiations depend upon their wavelength and frequency. To understand this, you need to know…

- the different types of electromagnetic radiation
- the uses and dangers of each type of radiation
- the difference between analogue and digital signals.

Electromagnetic Radiation

Electromagnetic radiations are disturbances in an electric field. They travel as **waves** and move energy from one place to another. Each type of electromagnetic radiation…

- has a different **wavelength** and a different **frequency**
- travels at the same speed (300 000 000m/s) through a vacuum (e.g. space).

Electromagnetic radiations (such as light) form a continuous range called the electromagnetic spectrum.

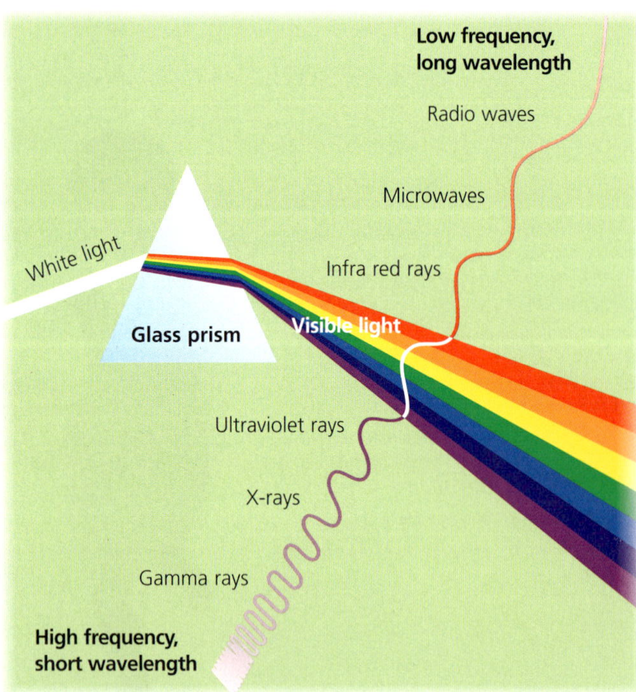

Radio waves, microwaves and infra red rays all have a longer wavelength and a lower frequency than visible light.

Ultraviolet waves, X-rays and gamma rays all have a shorter wavelength and a higher frequency than visible light.

Electromagnetic radiation can be **reflected** and **refracted**. Different wavelengths of electromagnetic radiation are reflected, absorbed or transmitted differently by different substances and types of surface, e.g. black surfaces are particularly good absorbers of infra red radiation.

When a wave is **absorbed** by a substance, the energy it carries is absorbed and makes the substance heat up. It may also create an **alternating current** of the same frequency as the radiation. This principle is used in television and radio aerials, which receive information via radio waves.

Electromagnetic waves obey the wave formula:

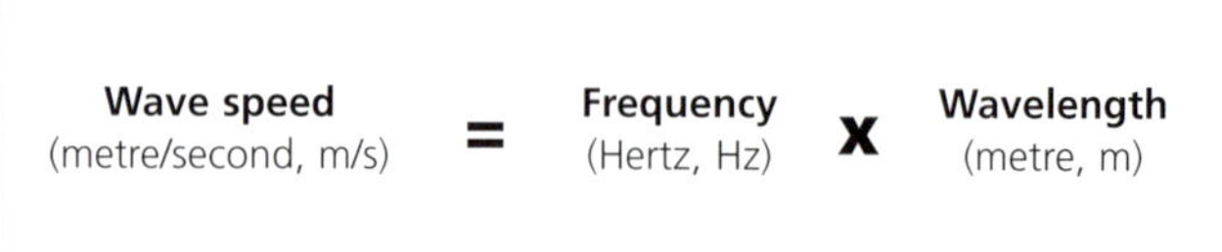

Visible Light

Light is one type of electromagnetic radiation which, together with the other various types of radiation, is in the electromagnetic spectrum.

The seven 'colours of the rainbow' form the **visible spectrum** which, as the name suggests, is the only part of the electromagnetic spectrum that we can see.

The visible spectrum is produced because white light is made up of many different colours. These are refracted by different amounts as they pass through a prism – red light is refracted the least and violet is refracted the most. Visible light can be used for communication through optical fibres, and for vision and photography.

Electromagnetic Waves	Uses	Effects
Radio Waves	• Transmitting radio and TV signals between places – waves with longer wavelengths are reflected by the ionosphere (an electrically charged layer in the atmosphere) so they can send signals between points regardless of the curve of the Earth's surface.	• Certain cancers can be treated with radio waves. • High levels of exposure for short periods can increase body temperature leading to tissue damage, especially to the eyes.
Microwaves	• Satellite communication networks and mobile phone networks. • Cooking – microwaves are absorbed by water molecules causing them to heat up.	• May damage or kill cells because they are absorbed by water in the cells, leading to the release of heat. Therefore, care must be taken in the use of microwaves.
Infra Red Rays	• Grills, toasters and radiant heaters. • Remote controls for televisions and video recorders. • Optical fibre communication.	• Absorbed by skin and felt as heat. • Excessive amount can cause burns.
Visible Light	See p.24	See p.24
Ultraviolet Rays	• Security coding – a surface coated with special paint absorbs UV and emits visible light. • Suntanning and sunbeds.	• Passes through skin to the tissues below. Darker skin allows less penetration and provides more protection. • High doses of this radiation can kill cells and a low dose can cause cancer.
X-rays	• Producing shadow pictures of bones and metals.	• Passes through soft tissues (although some is absorbed). • High doses can kill cells and a low dose can cause cancer.
Gamma Rays	• Killing cancerous cells. • Killing bacteria on food and surgical instruments.	• Passes through soft tissues – although some is absorbed. • High doses of this radiation can kill cells and a low dose can cause cancer.

Communication Using Electromagnetic Waves

Sound, e.g. speech or music, can be sent over long distances if it is converted into electrical signals which match the frequency and amplitude (the maximum disturbance caused) of the sound waves.

These electrical signals can then be sent using...

- **Cables** – copper cables weaken the signal during transmission. To boost the signal, regular amplification of the signal is required.
- **Electromagnetic waves** – a radio wave (called a 'carrier') is used to carry the electrical signal from a transmitter. This produces a **modulated** wave.

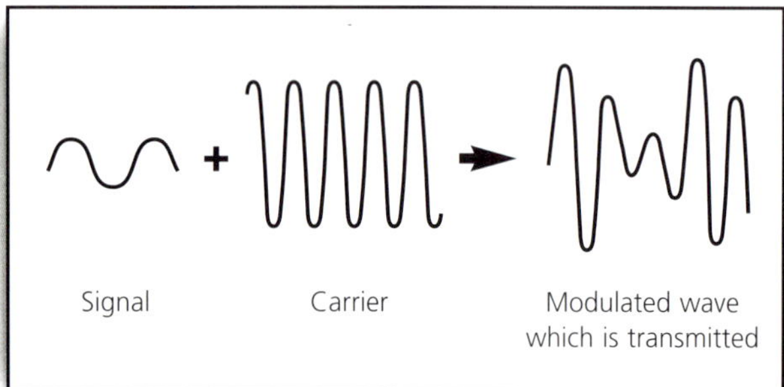

The modulated wave is picked up by the aerial in a radio and is demodulated (i.e. the carrier wave is removed) to leave the original signal. This is what comes out of a loudspeaker.

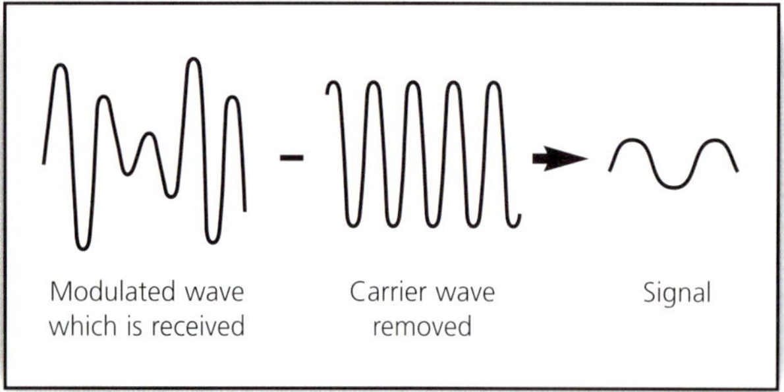

Optical Fibres

Information can also be transmitted using optical fibres where the electrical signal is converted into light or infra red pulses. These waves travel in curved paths.

More information can be sent this way than by sending electrical signals through cables of the same diameter, and there is less weakening of the signal along the way.

Analogue and Digital Signals

Analogue signals vary continually in amplitude and / or frequency. They are very similar to the sound waves of speech or music.

Digital signals do not vary. They have discrete values only, usually **on** (I) or **off** (0); there are no in-between states. The information is a series of pulses.

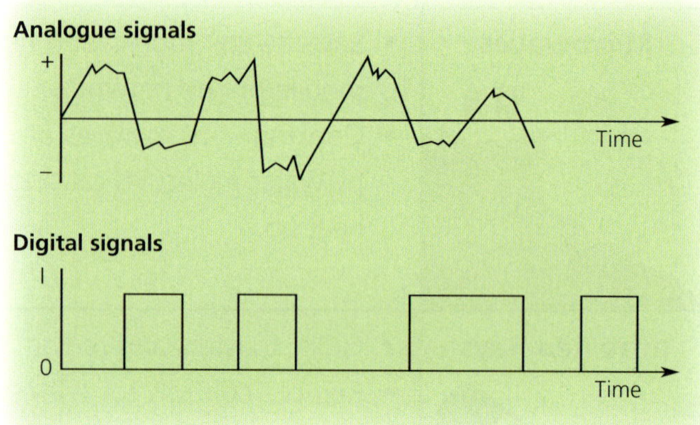

Digital Versus Analogue

Digital signals are of a better quality than analogue signals and there is no change in the signal information during transmission. More information can be transmitted in this way in a given time via cable, optical fibre or carrier wave. Digital signals can also be easily processed by computers.

When analogue signals are transmitted, they become weaker and can also pick up additional signals or noise (interference). The transmitted signals have to be amplified at selected intervals to counteract the effects of the weakening.

Different frequencies within analogue signals weaken by different amounts. During amplification, these differences and any background noise are also amplified, causing a deterioration in the signal quality.

Digital signals also weaken during transmission, but the pulses are still recognisable as 'on' or 'off'. The signals are unaffected by noise which usually has a low amplitude and is recognised as an 'off' state. When amplified, the quality of the digital signal is retained.

You need to be able to evaluate the possible hazards associated with the use of different types of electromagnetic radiation.

Example

It's Good to Talk – Or Is It?

New findings raise concerns that mobile phones could cause cancer and other health problems.

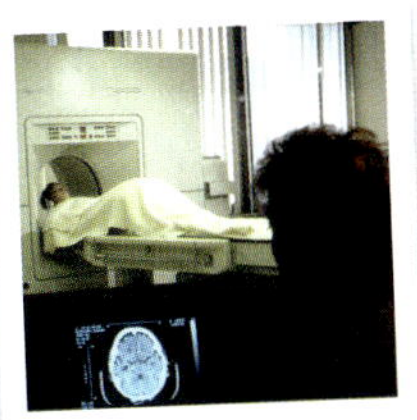

Swedish scientists studying the effects of electromagnetic radiation on red blood cells have found that levels of radiation, equivalent to those emitted by mobile phones, have a significant effect on the attractive forces between cells.

Up until now, the conventional view has been that radio waves could only cause damage at a cellular level if they carried enough energy to break chemical bonds or 'fry' tissue. These new findings might suggest that mobile phones do in fact emit enough energy to affect the bonds.

Experts, however, have been quick to point out that the results were obtained through tests on small groups of cells and provide no real evidence of a danger to health.

There have been other suggestions in the past that mobile phones can cause brain tumours and Alzheimer's disease, but so far research has been inconclusive.

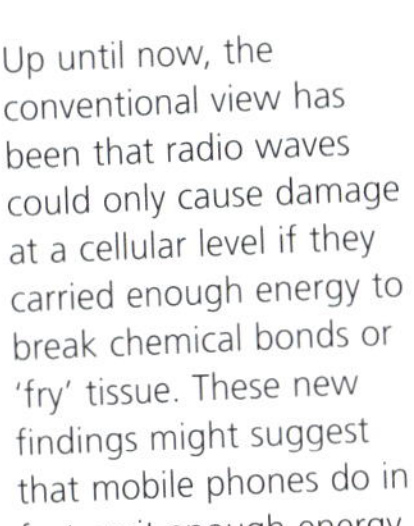

Mobile Phones are Safe

New study has found no link between mobile phones and cancer.

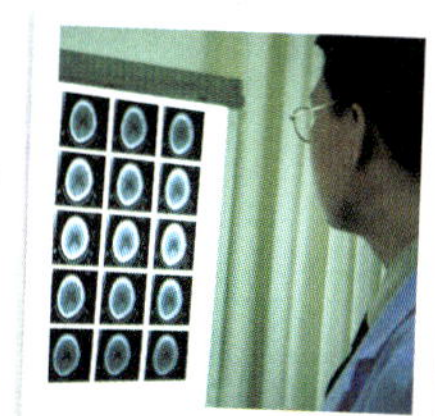

Scientists studying the possible links between mobile phones and brain tumours have reported today that they have found no correlation between using mobile phones and the risk of developing glioma – the most common type of brain tumour.

The study of 2 682 people across the UK looked at 966 individuals with diagnosed glioma and 1 716 individuals without the condition. It concluded that although a large percentage of the cancer sufferers reported their tumours to be on the side of the head where they held the phone, for regular mobile phone users, there was no increased risk of developing glioma.

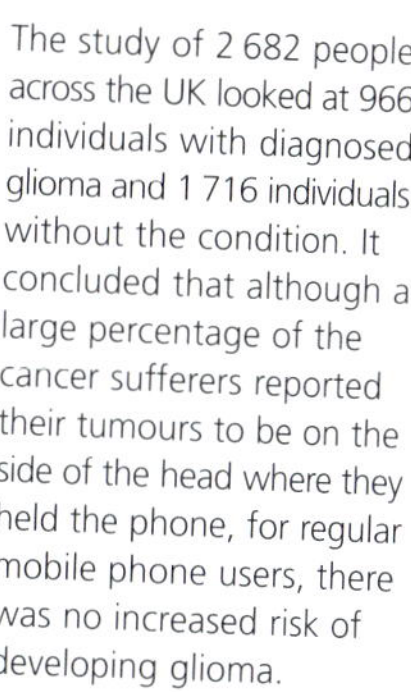

Using mobile phones

Benefits	Problems
• Easy, convenient method of communication, especially in a vulnerable situation – when car breaks down, alone at night, feel threatened, etc. • Can be used to access the Internet, take pictures, and watch television / video clips. • Easy way to keep in contact when away from home or abroad, e.g. text messages. • Many different tariffs and networks, which makes them affordable. • Can help in solving crime as mobile phones can be tracked.	• Some studies have linked mobile phone use with brain tumours and Alzheimer's disease. • The long-term effects of using mobile phones are not known – studies are still being carried out. • Increasingly, advertising is targeted at younger age groups who would be more vulnerable to any health implications.

How Science Works

You need to be able to evaluate methods to reduce exposure to different types of electromagnetic radiation.

Prolonged exposure to many types of electromagnetic radiation is considered harmful. There are ways in which exposure can be reduced.

Example

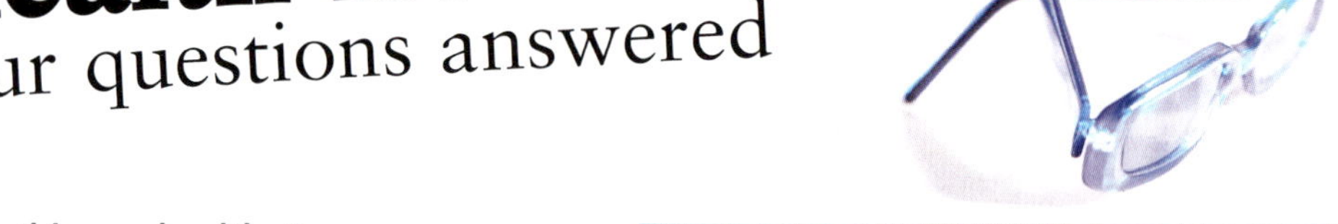

Health Issues

Your questions answered

Is sunbathing unhealthy?
I love sunbathing and going out in the Sun, but I have recently become concerned about the effects and dangers of exposure to the Sun. Could you please give me some advice as to the safest methods of dealing with the Sun and UV radiation?
Claire, aged 16, Southampton

As you know, some exposure to sunlight can be enjoyable. However, too much can be dangerous, causing immediate effects such as sunburn, blistering of the skin and heat stroke, whilst longer-term effects can include skin cancer, cataracts, wrinkling and ageing of the skin. Some scientists believe that too much UV radiation can even impair the immune system.

However, don't worry, there are some simple ways to ensure that you can enjoy the Sun and the benefits of UV light, whilst minimising the risks.

Wear sunglasses. Fashion sunglasses won't be enough though, you must get sunglasses that block 99–100% of UV light.

Wear a hat with a wide brim. A hat is very important to stop your brain from overheating which leads to sunstroke.

Protect other areas of your body with clothing. Make sure you choose light materials: anything too heavy and you could overheat.

Always use a high factor sun cream. For it to be effective you must reapply frequently, especially if you have been swimming.

Avoid the hottest parts of the day, between 11am and 3pm. I know this may seem restrictive, but the Sun is most intense, and most damaging, during these times.

Avoid sunlamps and tanning parlours. I know that these can often make you feel good, or help medical conditions such as SAD (seasonal affective disorder) or eczema, but they are damaging to your skin.

Keep hydrated: drink lots of water – this will help your skin to cope with the effects of the Sun better.

Apply lots of aftersun lotion to help protect your skin.

11.6

What are the uses and dangers of emissions from radioactive substances?

Radioactive substances constantly emit radiation from the nuclei of their atoms. Some nuclear radiations can be very useful but others can be dangerous. To understand nuclear radiation, you need to know...

- the basic structure of an atom
- what an isotope is
- the properties and uses of radioactive substances
- the meaning of 'half-life'
- the effect of radiation on living organisms.

Each atom has a small central **nucleus** made up of **protons** and **neutrons**. This nucleus is surrounded by **electrons**. Here is a helium atom.

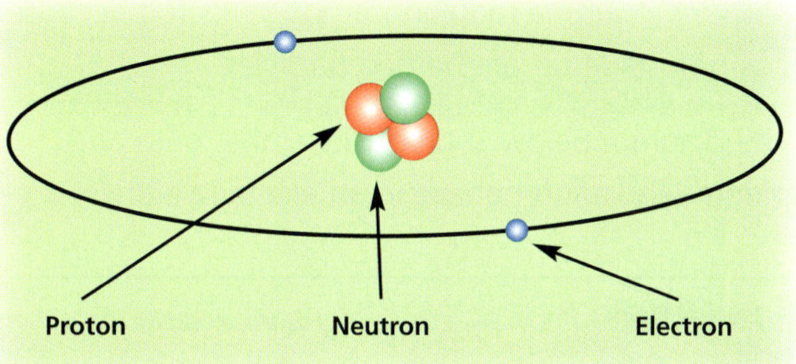

Isotopes

Although all the atoms of a particular element have the same number of protons, they can have a different number of neutrons. These are called **isotopes**.

Radiation

Some substances give out radiation all the time, regardless of what is done to them. These substances are said to be **radioactive**. Radiation is released from the nucleus as the result of a change in the structure of the atom.

There are three types of radiation:

- **alpha (α)** – an alpha particle is a helium nucleus (a particle made up of two protons and two neutrons)
- **Beta (β)** – a beta particle is a high-energy electron which is ejected from the nucleus
- **Gamma (γ)** – high-frequency electromagnetic radiation.

When radiation collides with atoms or molecules, it can knock electrons out of their structure, creating a charged particle called an **ion**.

The relative ionising power of each type of radiation is different, as is their power to penetrate different materials and their range in air.

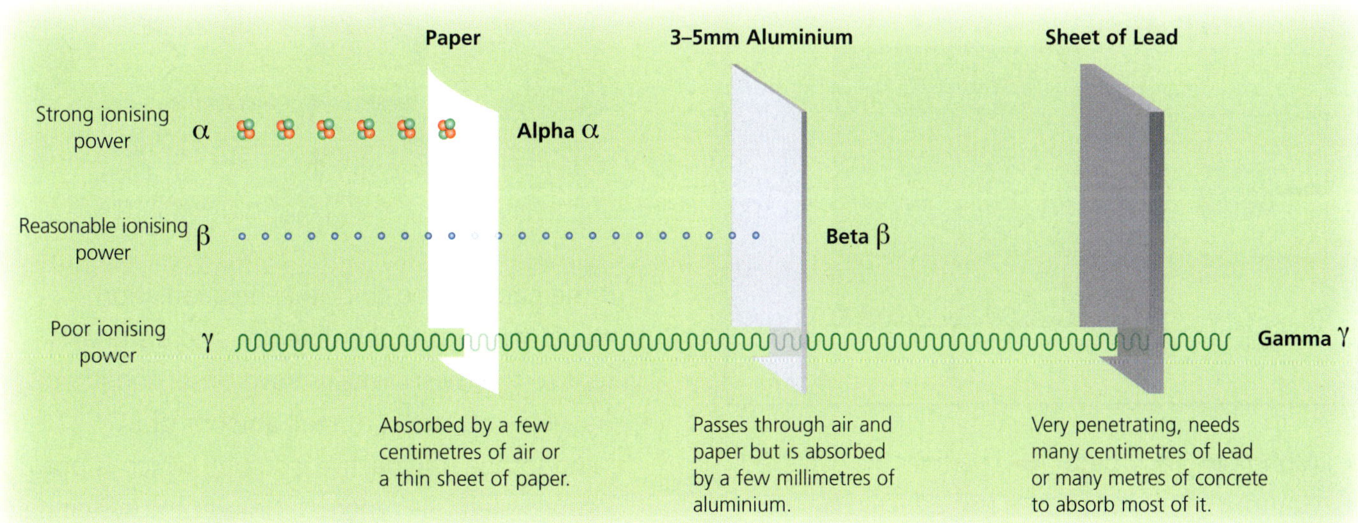

Electric and Magnetic Fields

Alpha and beta radiation are deflected by electric and magnetic fields. This is because they are charged particles. Gamma radiation is not deflected because it is not made up of charged particles.

The diagram below shows how alpha, beta and gamma radiation behave in an electric field:

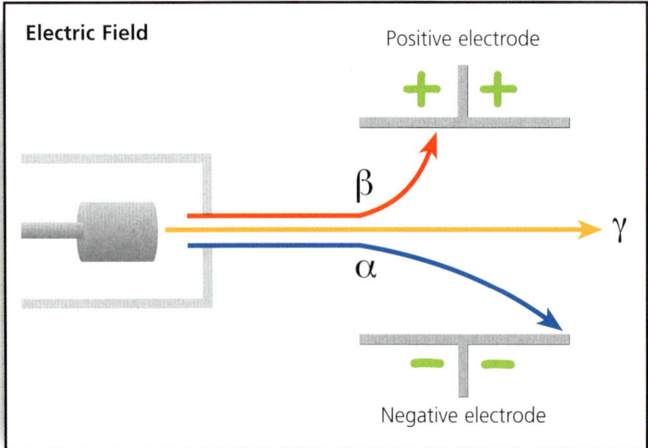

- **Alpha** (➤) particles are positively charged and are deflected towards the negative electrode.
- **Beta** (➤) particles are negatively charged and are deflected towards the positive electrode.
- **Gamma** (➤) radiation is not deflected by the electric field.

The diagram below shows how alpha, beta and gamma radiation behave in a magnetic field:

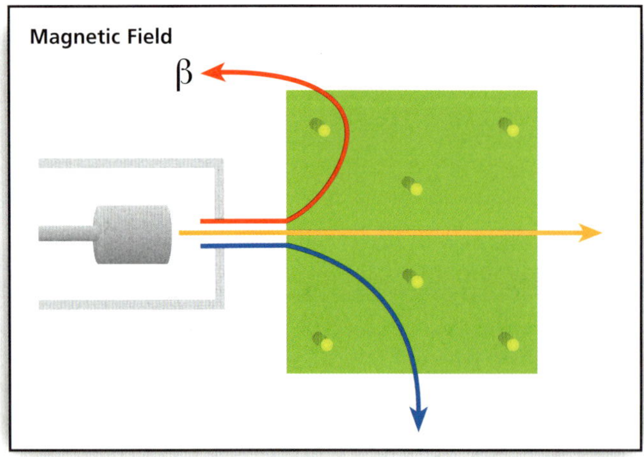

- **Alpha** (➤) and **beta** (➤) particles are deflected.
- **Gamma** (➤) radiation is not affected by the magnetic field.

Common Uses of Radiation

- Sterilisation – gamma rays can be used to sterilise medical instruments and food because they destroy germs and bacteria.
- Treating cancer – gamma radiation can be used to destroy cancerous cells. A high calculated dose is used from different angles so that only the cancerous cells are destroyed.

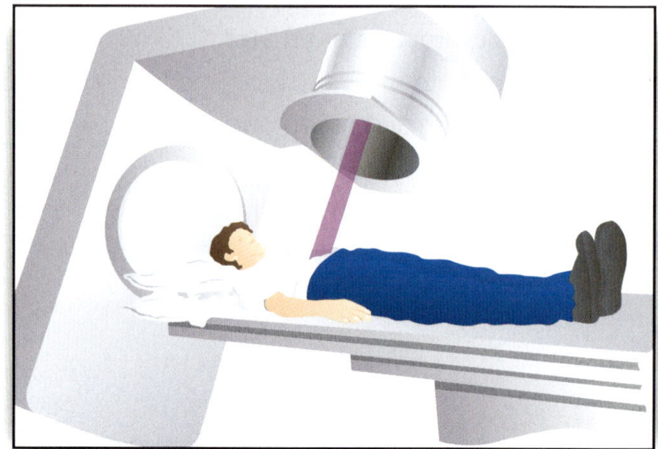

- Controlling the thickness of materials – when radiation passes through a material some of it is absorbed. The greater the thickness of the material, the greater the absorption of radiation. This can be used to control the thickness of different manufactured materials, e.g. paper production at a paper mill (see below).

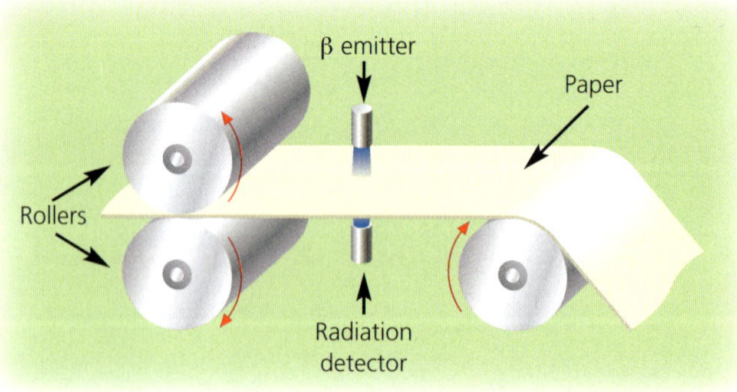

If the paper is too thick then less radiation passes through to the detector, and a signal is sent to the rollers which move closer together.

- Tracers – a tracer is a small amount of a radioisotope (radioactive isotope), which is put into a system. Its progress through the system can be traced using a radiation detector.

Acute Dangers

The ionising power of alpha, beta and gamma radiation can damage molecules inside healthy living cells, which results in the death of the cell.

Damage to cells in organs can cause cancer. The larger the dose of radiation the greater the risk of cancer.

The damaging effect of radiation depends on whether the source is located inside or outside the body. If the source is outside the body...

- α cannot penetrate the body and is stopped by the skin
- β and γ can penetrate the body to reach the cells of organs and be absorbed by them.

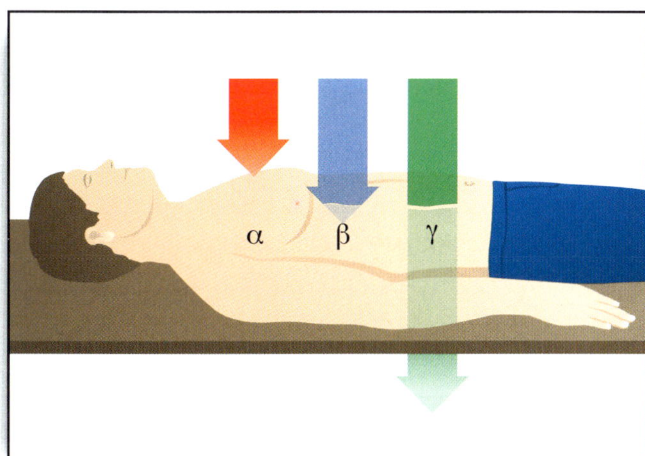

If the source is inside the body...

- α causes most damage as it is easily absorbed by cells causing the most ionisation
- β and γ cause less damage as they are less likely to be absorbed by cells.

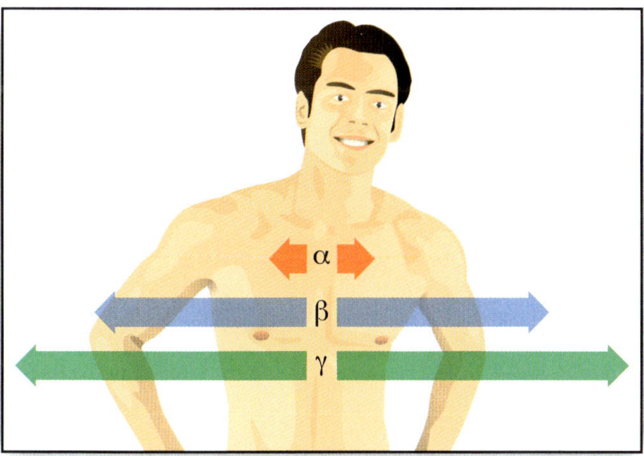

Half-life

The **half-life** of a radioactive isotope is a measurement of the rate of radioactive decay. It is the time it takes for...

1. the number of nuclei of the isotope (those which have not decayed) in a sample to halve

or

2. the count rate (the number of atoms which decay in a certain time) of a sample containing the isotope to halve.

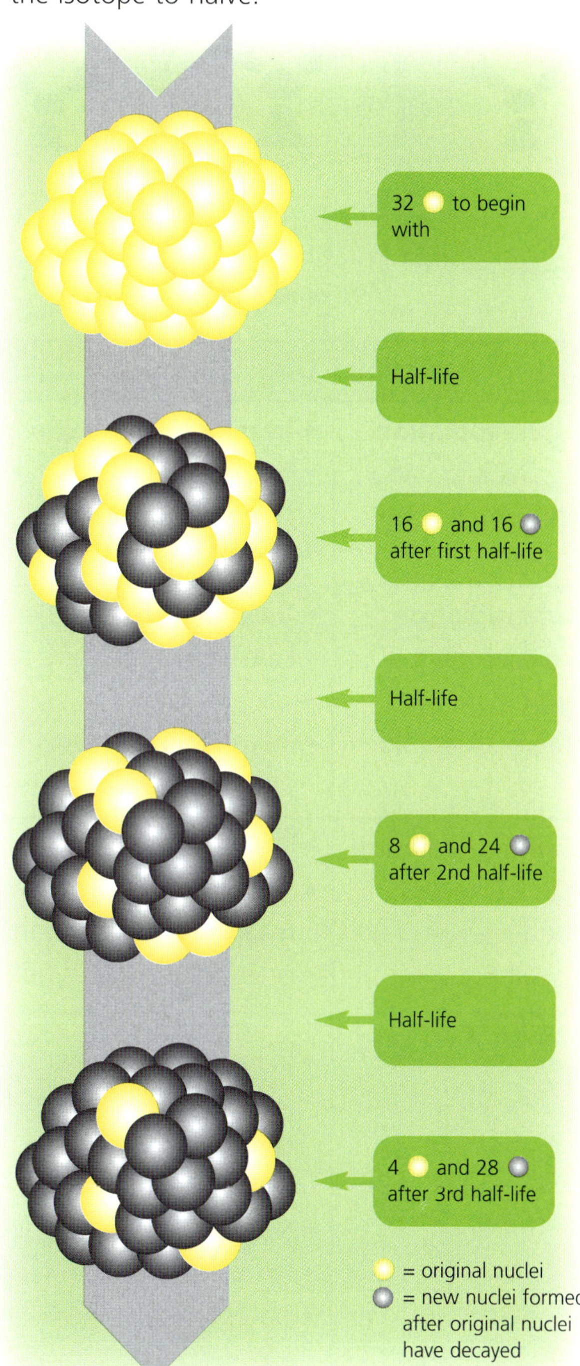

How Science Works

You need to be able to evaluate the possible hazards associated with the use of different types of nuclear radiation.

The risks to humans from radiation differ depending on the type of radiation and the source. We need to understand these risks so that appropriate safety precautions can be taken.

Type of Radiation	Uses	Dangers
Alpha radiation e.g. radium, radon, uranium, thorium.	• Treating cancer, e.g. mouth cancer. • Smoke alarms.	• Cannot penetrate the skin, but harmful if materials emitting alpha radiation enter the body, e.g. are inhaled or swallowed. • Difficult to detect through other materials.
Beta radiation e.g. strontium-90, carbon-14, tritium, sulfur-35.	• Treatment of thyroid disorders. • Luminous signs, dials and gauges. • Carbon dating. • Industrial instruments for gauging thickness of materials (e.g. in paper mills).	• Can penetrate human skin cells and cause skin damage. • Materials emitting beta radiation may be harmful if they enter the body.
Gamma radiation e.g. iodine-131, caesium-137, cobalt-60, radium-226.	• Treating cancer. • Industrial control systems which measure and control the flow of liquids or ensure packages (of food, drugs, etc.) are filled to the correct level. • Industrial instruments, e.g. in geological surveys, assessing sites for mining and construction and gauging thickness of materials in steel mills. • Sterilising medical equipment. • Pasteurising certain foods and spices. • Tracers (used in medicine and agriculture).	• Very penetrating, so even sealed sources can be a hazard. • Can easily penetrate the body and damage / kill cells in the internal organs. • Can cause cancer. • High doses can kill.

You need to be able to evaluate measures that can be taken to reduce exposure to nuclear radiation.

The transportation and application of radioactive substances is carefully controlled by government rules and regulations to minimise the risks to the general public. Authorised personnel who handle radioactive substances or operate machinery follow strict guidelines. These vary depending on the risk factor involved. The table below lists some measures that can be taken to reduce / prevent exposure to different types of radiation.

External Exposure – Beta and Gamma

- Minimise time of exposure.
- Maximise distance from the source of radiation.
- Wear protective clothing (and remove it before leaving a restricted area).
- Avoid direct handling of radioactive materials, i.e. use implements like forceps and tongs.
- Use protective shields, screens and containers.
- Use instruments that can detect levels of radiation, e.g. Geiger-Müller counter or liquid scintillation counter.
- Use materials which can provide a shield against radiation, e.g. lead, concrete and water.
- Wear a film badge which monitors the degree of exposure to radiation (see below right).

Internal Exposure – Alpha, Beta and Gamma

- Wear chemical fume hoods and protective masks to prevent inhalation.
- Never consume or store food and drink close to a radioactive source.
- Wear protective clothing.
- Ensure that any cuts or wounds are sealed up.
- Minimise amount of radioactive material to be handled.

People who work in the nuclear industry and are regularly exposed to radiation often wear a badge to monitor their degree of exposure. The badge contains photographic film, which (after developing) becomes darker the more radiation it has been exposed to. In this way, radiation exposure can be carefully monitored.

How Science Works

You need to be able to evaluate the appropriateness of radioactive sources for particular uses, including as tracers, in terms of the types of radiation emitted and their half-lives.

The properties of different radioactive substances make them suited to different uses.

Example 1 – Medical Tracers
Doctors use radioactive chemicals called tracers to help detect damage to the internal organs. Once the tracer enters the body, it builds up in the damaged or diseased part of the body. Radiation detectors can then be used to detect where the problem is. These can be linked to computers which produce an image showing the distribution of the radioactive chemical. Problem areas are highlighted by a high concentration.

Because it is being used inside the body, the radioactive tracer must be non-toxic. It also needs to have a short half-life, so that it breaks down quickly after use. Gamma and beta sources are used because they pass out of the body easily. An alpha source is never used because it would be quickly absorbed and cause damage.

Technium-99 is a gamma emitter often used for this purpose. It has a half-life of 6 hours. This gives the doctors enough time to detect the problem, but ensures that the chemical does not stay in the body for too long.

Example 2 – Industrial Tracers
Any leaks in a pipeline can be found by injecting a gamma-emitting isotope into the system. The tracer will leak out into the soil where the pipe is broken. Because of the penetrating nature of gamma rays, the leak will be easy to detect through several feet of soil.

A short half-life radioisotope is used so that it does not remain in the environment any longer than is necessary after the leak has been detected.

Example 3 – Controlling the Thickness of Materials
Radiation is absorbed by material through which it passes. This effect can be used to monitor the thickness of materials in production. Beta rays can be used to control the thickness of paper, and gamma rays can be used to control the thickness of metal.

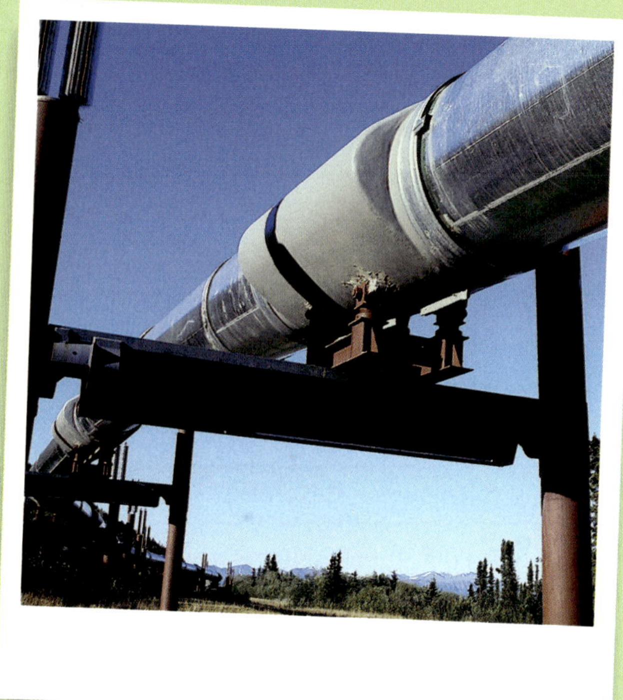

Radiation	Use	Suitability
Gamma	Medical and industrial tracers.	Can pass out of the body or pipeline easily. Short half-life so does not stay in the body or environment for long.
	Controlling the thickness of metal.	Passes through thin metals, but is absorbed by heavier metals, e.g. lead.
Beta	Medical tracers.	Easily transferred out of the body.
	Controlling the thickness of paper.	Passes through paper but is absorbed by thin metals.
Alpha	Medical tracers.	Unable to transfer out of the body, would be quickly absorbed and cause damage – never used.
	Controlling the thickness of paper.	Absorbed by paper, so cannot be detected.

11.7

What do we know about the origins of the Universe and how it continues to change?

Current evidence suggests that the Universe started from a 'big bang' at a small point, from which mass and space expanded violently and rapidly, and is still expanding. To understand this, you need to know...

- how we study the Universe
- about red shift and how it supports the 'big bang' theory.

Observing the Universe

Observations of the Universe can be carried out from space or on Earth. One method of observation is to use a telescope.

Different types of telescope can detect visible light or other electromagnetic radiations, e.g. radio waves or X-rays.

Reflecting Telescopes

This type of telescope reflects light using mirrors (see p.81).

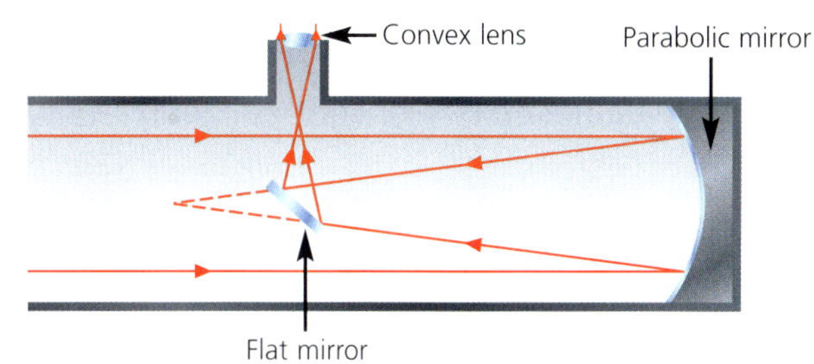

Refracting Telescopes

This telescope refracts light at each end using lenses (see p.81–83).

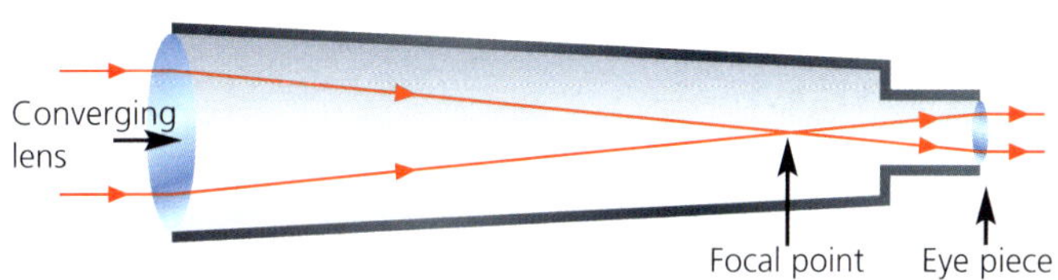

Radio Telescopes

As the name suggests, these telescopes pick up radio waves instead of light waves. Radio waves are much longer than light waves, so to be able to receive good signals, radio telescopes need large antennas or arrays of smaller antennas working together.

The radio waves are emitted by bodies in space. Most radio telescopes use a parabolic dish to reflect the radio waves to a receiver, which detects and amplifies the signal.

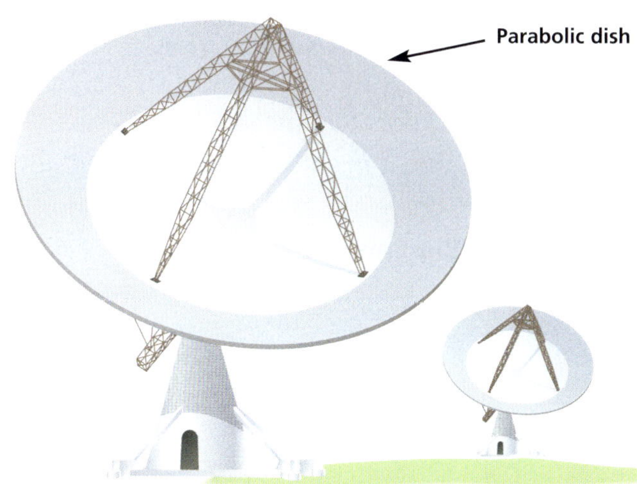

Using Telescopes

On Earth

Observing the Universe from the Earth is limited by the atmosphere. Interference from the atmosphere, clouds, weather storms or light pollution reduces the quality of images. Many telescopes are therefore placed on the top of mountains and in areas with low levels of pollution to reduce interference.

In Space

Putting telescopes into space so that they orbit the Earth outside the atmosphere means the images produced are not affected by atmospheric interference, e.g. the Hubble Telescope (see p.37).

Red Shift

If a wave source is moving away from, or towards, an observer there will be a change in the observed wavelength and frequency.

If a source of light moves away from us, the wavelengths of the light in its spectrum are longer than if it was not moving. This is known as **red shift** because the wavelengths 'shift' towards the red end of the spectrum (see p.24).

There is a red shift in light observed from most distant galaxies, which means that they are moving away from us very quickly (see picture 1).

This effect is exaggerated in galaxies which are further away, which means that the further away a galaxy is, the faster it is moving away from us (see picture 2).

This evidence suggests that the whole Universe is expanding and that it might have started billions of years ago, from one place with a huge explosion, known as the '**Big Bang**' (see picture 3).

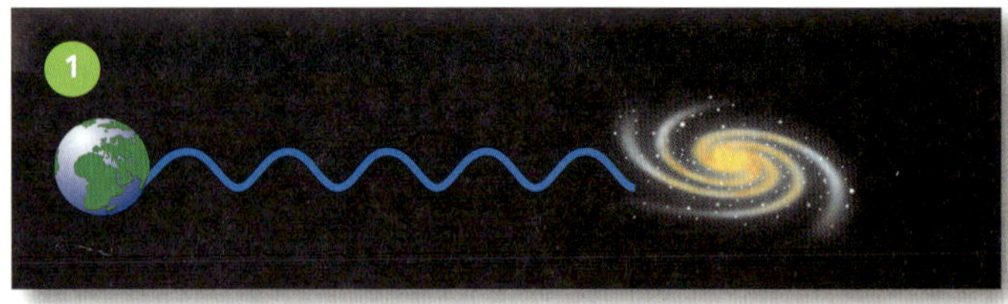

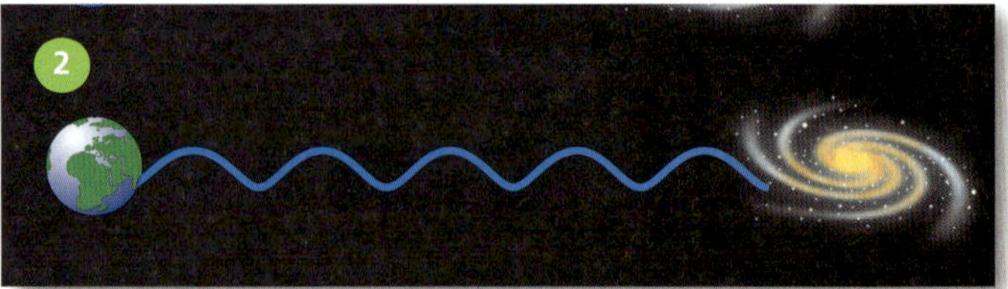